FISHERIES MANAGEMENT IN CRISIS

FISHERIES MANAGEMENT IN CRISIS

By

N.K. Bose

2008

CYBER TECH PUBLICATIONS
4264/3, Ansari Road, Daryaganj, New Delhi-110002 (India)
Ph.: 011-23244078. Fax: 011-23280028
E-Mail: cyberpublicationsdelhi@yahoo.com

FISHERIES MANAGEMENT IN CRISIS

N.K. Bose

First Edition 2008

Published by :

G.S. Rawat for **Cyber Tech Publications**
4264/3, Ansari Road, Daryaganj, New Delhi-110002 (India)
Ph.: 011-23244078. Fax: 011-23280028
E-Mail: cyberpublicationsdelhi@yahoo.com

Printed at : **Amit Enterprises**, Maujpur, Delhi, (9810878184)

Preface

The seeds of expansion beyond the modeling approaches were the many aspects of fish and of fisheries that had been left on the sideline over the decades of the development of the three models. Those seeds included that few fisheries were focused on only one species or population, and that, even when they were, fishing affected other organisms in the ecosystem, and indeed the environment itself. Further, fisheries biology had explicitly excluded accounting for fishing being both an economic and a human enterprise. Management advice that did not also account for these factors was at best incomplete and, at worst, left many fisheries by the beginning of the 21st century either collapsed or vulnerable to collapse. The science necessary for the management of fisheries is something much broader than fisheries biology.

Taking a perspective that individual fisheries are in fact the objects of study by fisheries science, we suggested that such autopsies would involve conducting comprehensive examinations of individual fisheries aimed at determining the success (or not) of fisheries science and of fisheries management.

Here I sincerely hope that this work of mine would be beneficial to the readers.

Suggestions and complaints are always welcome as they make us more working and correct. So please free to write us.

—Author

Contents

1

Systems Approach to Aquaculture Management

The topics covered in the Conference on Aquaculture in the Third Millennium, from highly technical sessions on genetics and feeds, to non technical policy sessions on the role of different stakeholders and institutions, reflect the wide range of factors and issues relevant to modern-day aquaculture development. The "systems approach" recognizes the diverse factors affecting aquaculture and is a multifactorial and multidisciplinary approach that attempts to analyse how different factors affect aquaculture and develop solutions to problems based on an understanding of how aquaculture systems operate. The systems approach is fundamentally a multi-disciplinary approach that can be used to solve problems and identify opportunities for development.

This analytical approach has been shown to contribute to identification of key research issues, development of better management solutions, improvement of business efficiency, design and testing of new aquaculture systems and more effective extension and education, among others. This paper discusses the systems approach to aquaculture, provides examples of the relevance and use of the approach in aquaculture development, and recommends areas for further study and follow up actions.

Introductory Remarks

The varied topics covered in these Proceedings of the Conference on Aquaculture in the Third Millennium, from highly technical sessions on genetics and feeds, to nontechnical policy sessions on the role of different stakeholders and institutions, reflect the wide range of factors and issues relevant to modern aquaculture development.

The "systems approach" recognizes this diversity of influences on aquaculture development, and is a multifactorial and multidisciplinary approach. It uses an understanding of how aquaculture systems operate to analyse how different factors affect aquaculture and develop solutions to problems that are identified. This analytical approach has been shown to contribute to identification of key researchable issues (Edwards, 1998; Smith, 1999), development of better management solutions, improvement of business efficiency, and design and testing of new aquaculture systems, as well as to more effective extension and education.

The analysis and understanding of aquaculture systems can occur at different "levels":

* the organism and its surroundings;
* the production unit;
* the economic enterprise;
* the farm, watershed or coastal areas;
* the national sectoral level, or even
* the international level.

The boundaries chosen for the system of interest may be physical entities, political borders or organizational structures. In terms of this paper, the focus is on the following systems: the farm; the farm and its local environment; and the national level, recognizing the fact that there are critical interactions between these different levels.

This paper is intended to be complementary to the earlier papers in this technology session which focused on the individual components of the system-such as feed and nutrition, health, seed and genetics-and emphasises understanding of how these individual components interact within the wider context of aquaculture systems. It first looks at our current understanding of aquaculture systems and management strategies based on a systems approach, with an emphasis (as in this session) on farm-level management practices.

As social, economic, policy and institutional issues are all important-at some level-in a systems approach, the paper also provides a link between the Conference's farming technology and policy sessions and closes with some suggestions for discussion on future directions. This paper was developed from discussions and contributions made by panel members and participants during the Conference on Aquaculture in the Third Millennium. The recommendations presented here represent a consensus of recommendations adopted in the final plenary session of the Conference.

AQUACULTURE SYSTEMS AND SUSTAINABILITY

Sustainable development and sustainability are complex issues that are difficult to define and apply to aquaculture. The "systems approach", however, can assist understanding of these issues, as they relate to aquaculture development.

The term sustainability has been defined in various ways but perhaps the most widely used is based on the definition of "sustainable development" in the Brundtland report: "sustainable development is development that meets the needs of the present without compromising the ability of future generations to meet their own needs". An even more succinct definition is that of the International Union for the Conservation of Nature (IUCN): "sustainable development improves people's quality of life within the context of the Earth's carrying capacity". These definitions contain two key concepts: meeting the present and future needs of the world's people, and accepting the limitations of the environment to provide resources and to receive wastes for the present and for the future. The Food and Agriculture Organization of the United Nations (FAO), in particular, recognized that increased capacity at the national level is required to achieve sustainable development by including the need for "institutional change" in definitions of sustainable agricultural development (FAO, 1995). The recognition that institutions are important highlights the need for education and training, effective institutional arrangements and a legal and policy framework to underpin sustainable development of agriculture, and indeed aquaculture.

Sustainability is commonly split into three separate components: social sustainability (SS); economic sustainability (EcS); environmental sustainability (ES). Whilst social sustainability criteria are difficult to define, the definition of economic and environmental sustainability are providing a basis from which management options can evolve in aquaculture projects. To take sustainability to a more practical level requires consideration of environmental, social and economic issues in aquaculture development. Thus, the approach to sustainability implies a systems approach.

There are general guidelines available on the different issues to consider. The Code of Conduct on Responsible Fisheries (CCRF), adopted by the FAO Conference in

1995 (FAO, 1995) in particular, identifies a number of key issues. The Code sets out principles and international standards of behaviour for responsible practices, to ensure effective conservation, management and development of living aquatic resources while, at the same time, recognizing the nutritional, economic, social, environmental and cultural importance of fisheries and aquaculture, and the interests of all those involved in these sectors. The Code's Article on Aquaculture Development (Article 9) contains provisions relating to aquaculture, including culture-based fisheries. Fundamentally, the Code recognizes the importance of activities that support the development of aquaculture at different levels:

* the producer level;
* the local area, i.e., the farm and its integration into local area management and rural development schemes;
* the national institutional and policy environment; and
* international and transboundary issues.

The Code identifies many key principles in development of management strategies based on an understanding of aquaculture systems-from the farm to national and international levels. It also provides a basis for a systems approach.

THE SYSTEMS APPROACH

Farm Level

There is a lot of information on aquaculture farming systems, and various definitions are available, such as the level of intensity of management and output, and degree of integration with other on-farm activities. A considerable literature exists on integrated (agriculture-aquaculture and vice versa) farming systems (Edwards,

1998 and "Farming species and systems" in these proceedings).

However, there are a wide range of culture species, culture facilities and management practices in use, and thus a very wide range of farming systems.

Key factors to be understood in the functioning of a farming system are the technologies of production and social, economic and environmental aspects. At the technology level, feeds, feed additives and fertilizers, water quality, seed quality and availability, chemoterapeutants and other chemicals, disposal of wastes that may adversely affect human health and/or the environment, and food safety of aquaculture products all require consideration. It has also been emphasised that a better understanding of the microbial populations in aquaculture systems, their interaction with the health of the farmed animals, and their role in maintaining a healthy aquatic environment are required.

The systems approach at the farm level can be used to understand and improve the efficiency of use of key natural resources-particularly water, nutrients, land, seed and financial resources. When focussing on the mix of all sustainability criteria through a systems approach, we have to deal with such factors as: appropriate densities, production and husbandry systems geared to the animal's health, maintaining ecological balance within the pond or other growout habitat, and provision of optimal social and economic benefits. It is inevitable that impacts on natural resources will become an increasingly important issue in the new millennium, and the systems approach can be used to analyse and develop the solutions required for more sustainable use of natural resources.

A wide range of management systems is already employed in aquaculture operations with varying degrees of success. Given that aquaculture systems range from small, relatively self-contained farms for subsistence, to large-scale commercial units for trade purposes, variable success is hardly surprising. Thus, a "one-size fits all" approach is unlikely to be successful. A systematic approach to production management, however, allows the farmer to manipulate and control production inputs that will result in more efficient, cost-effective production and minimize excessive outputs with negative environmental impacts. There is a tremendous body of information on site selection, farm construction and design features, aquatic animal health management, broodstock and seed production and care, production techniques, the use of appropriate feeds, feed additives and fertilizers, water and sediment management, including effluent control, and other topics (e.g., Chanratchakool et al. 1998; FAO/NACA, 1995; Pennell and Barton, 1996).

The challenge is to optimize dissemination and use of such information and experience.

The systems approach can also be used for aquatic animal health management. Here the emphasis needs to move more towards management procedures, policies and products that can prevent or effectively eradicate significant pathogens, prevent re-infection through contaminants, and manage diseases in an environmentally sustainable manner. Subasinghe et al. (1998) provide a discussion of the role of the systems approach in aquatic animal health management.

The Local Level

The systems approach at the local level recognizes that individual farms cannot be seen in isolation, and that there are many interactions between an aquaculture farm and the external environment-including environmental resources and local communities.

Furthermore, there can be significant cumulative effects where there are large numbers of farms crowded in small areas.

Environmental interactions with aquaculture arise from a wide range of inter related factors including availability, amount and quality of resources; type of species cultured; size of farm; culture systems management; and environmental characteristics of the farm location. Environmental interactions are not limited to impacts of aquaculture on the environment, but include environmental impacts on aquaculture and impacts of aquaculture on aquaculture. Perhaps less well known or documented are the many ways that aquaculture can contribute to environmental improvement, for example mollusc farms' desedimentation or improved nutrient turnover (Hatcher et al., 1994), or water storage on small-scale freshwater farms.

At the local level, social and institutional interactions are also important and need to be better understood, for example:

* participation of, and benefits to, rural communities;
* institutional support through extension services;
* access to information etc.

A systems approach attempts to understand these linkages and develop management strategies based on such understanding.

The future of integration of aquaculture into local ecological and social systems requires more focus on local area development planning. Fortunately, increasing attention is being given to such issues, particularly in coastal areas. Integrated coastal management (ICM) is a process that addresses the use, sustainable development and protection of coastal areas, and according to GESAMP (1996) "comprehensive area-specific marine management and planning is essential for maintaining the long-term ecological integrity and productivity and economic benefit of coastal regions". ICM is made operational through such activities as:

* land use zoning and buffer zones;
* regulations, including permitting to undertake different activities;
* nonregulatory mechanisms;
* construction of infrastructure;
* conflict resolution procedures;
* voluntary monitoring; and
* impact assessment techniques.

More participatory approaches to planning of aquaculture development will also be given attention with the move towards integrated development planning (see S. Sen, this volume). Practical experience in implementation ICM for aquaculture is limited, which is in large measure because of the absence of adequate policies and legislation and institutional problems, such as a lack of unitary authorities with sufficiently broad powers and responsibilities, as well as limited training and education of people concerned.

In inland rural areas, increasing attention is being given to integration of aquaculture into rural development and special area management plans. Increasing emphasis is also being given to promotion of aquaculture for poverty alleviation (see Haylor and Bland, this volume). Such an approach requires emphasis on immediate social needs and people's livelihoods (and how aquaculture might meet these needs and contribute to improved livelihoods) rather than a technology/aquaculture driven approach. The emphasis on aquaculture for development, rather than development of aquaculture may lead to some fundamental changes in the approach to promotion of aquaculture in the coming years.

National, Regional and International Levels

At the national level, government policy, and institutional and human capacity are most important in providing a strong foundation for aquaculture to develop in a su (Insull and Shehadeh, 1996).

These issues are covered extensively in the Conference's policy session and are not discussed in detail here, except by recognition that most community and farm activities are influenced by national-level policy, legislation and institutional support. For example, the level of aquatic animal disease affecting small-scale producers or enhanced fisheries is related to national policies for quarantine and movement of live aquatic animals, which affect the risk of exposure of small-scale producers to serious aquatic pathogens (DFID/FAO/ NACA/GoB, 2001). Inter-and intra-country trading patterns and movement of aquatic animals also affect these risks. International conventions (e.g. the Convention on Biodiversity), trade and consumer preferences, all clearly impact aquaculture development at a local level.

APPLICATION OF THE SYSTEMS APPROACH

Some examples of systems approaches presented during the Conference are given below.

Small-scale Farmer Research and Extension

The systems approach has been widely used in the promotion of sustainable development through small-scale integrated aquaculture (Edwards, 1998). He emphasises that most scientists focus on technical aspects of aquaculture, resulting in the impression that the major constraint facing aquaculture development is a shortage of technical knowledge, overshadowing the developmental and educational constraints. The most important constraint to aquaculture development is dissemination of existing knowledge, whether derived from research or indigenous technical knowledge of farmers also affect these risks. The limited capacity of developing-country institutions in education, research and development compounds this fundamental failing. Research should follow farming systems research and extension methods in which inter-disciplinary teams work with farmers to evaluate and develop both production systems and extension methods that are appropriate to the local conditions of farmers and their resource base.

A systems approach to management has more potential for success on the typically diverse small-scale farms in developing countries. This approach includes analysis of the resource-base of the farm, and the farmers' perceptions of their needs. The role of institutions in promoting the development of aquaculture can also be analysed using such an approach. The role of national institutions (government and non governmental agencies) as primary facilitators of aquaculture and the necessity of capacity building in order to facilitate the systems development approach should be recognized. International and regional agencies can support such capacity building.

ENVIRONMENTAL MANAGEMENT SYSTEMS

Environmental management systems are also starting to evolve for some forms of aquaculture, as well as more formal environmental management systems (EMS). EMS is a complex approach with little direct application to small-and medium-sized farms, however, EMS principles provide useful guidance for improving environmental management of aquaculture production systems (H. Dixon, pers. comm.).

EMS assemble management policies, programmes and practices designed to identify links between industry, urban and developmental activities, and consequent pressures on the environment. An effective EMS for aquaculture should establish indicators of changes in the environment, including land, water and aquatic resources. Policies and practices responding to the changes are implemented with continuous feedback to reduce/mitigate any environmental impacts. These indicators should make clear links between environmental impacts and aquaculture activities. The indicators should also reflect positive and negative impacts of environmental influences on aquaculture, as well as visa versa. EMS provides a possible systematic approach to motivate aquaculture to better organize priorities and projects to identify problems and potential impacts before they occur, as well as meet environmental and business goals. This process also assists compliance with national environmental laws and regulations. A successful EMS provides the means by which aquaculture can identify causes of environmental problems and prevent them, thus saving money to repair or mitigate after the damage has been done.

According to H. Dixon, key benefits of an EMS include:

* improved environmental performance,
* reduced liability,
* competitive advantages,
* improved compliance,
* reduced operational costs,
* enhanced consumer trust, and
* increased access to capital.

H. Dixon considers that formulating an institutional framework for environmental management systems of aquaculture should:

* be simple to implement, clear and comprehensible to all involved, including the public;
* consider the needs of all stakeholders;
* consider individual abilities and resources;
* be financially flexible and not inhibit the activities it is designed to address; and
* be based on sound scientific information, quantifiable and effective.

CODES OF PRACTICE

The technical methods, management systems and practices needed for minimizing impacts are being increasingly incorporated into more formal "Codes of Practice", notably in more commercially oriented and intensive shrimp and salmon farming. Several aquaculture organizations, for example, the Marine Shrimp Culture Association of Thailand, the Global Aquaculture Alliance, the Australian Prawn Producers Association, the Irish Salmon Growers Associations and others have taken the FAO Code of Conduct a step further and formulated Codes of Practice (COP). These COP contain principles for preventing or mitigating negative environmental and social impacts through use of "best management practices" (BMP).

They are currently for voluntary adoption and consist of documented guidelines available to farmers. Much needs to be done towards their implementation, such as the development of operational manuals and support programmes. Furthermore, the extent to which COP will be fully adopted by farmers under self-regulation and the environmental consequences of their adoption remain to be determined. Implementation is a very important issue for the new millennium, particularly for small-scale farmers.

Improving Profitability of Aquaculture Operations

The systems approach is directly applicable to the business of aquaculture, through the use of a structured, systematic approach to operations and business management.

E. Hempel considers that in high technology salmonid aquaculture, the "software" of aquaculture (people) is more important than the hardware or "technology". There are many facts and variables, so in order to understand and manage the system, things have to be simplified- "systematized"-. The approach emphasised by E. Hempel involves:

* understanding and identifying the system of concern within the farm enterprise/business;
* simplifying the system by identifying only key system variables;
* for each person working at different levels of the organization, identifying the key system variables to monitor;
* reducing the number of variables to the minimum- e.g. dissolved oxygen, food conversion ratio (FCR), cash flow, and set limits/standards; and
* monitoring and responding to changes that exceed the set limits.

E. Hempel also emphasises that effective management based on a systems approach requires assigning of responsibilities within an aquaculture business to monitor and respond to changes that occur. He also emphasises that the system may change with time, and that management has to be flexible to respond to such changes. He further considers that the systems approach to business and management is important to improving the performance and profitability of farming, and is relevant at all levels of aquaculture, from small-scale farms to the largest business.

Institutional use of a Aystems Approach

A systems approach can also be used to define institutional responsibilities in aquaculture. For example, in western Australia, according to C. P. Rogers' presentation during the "Systems session" of the Conference on Aquaculture in the Third Millennium, NACA/FAO, 2000) a systematic analysis of institutional responsibilities was useful to establish the key points for decision making in allocation of land and water resources for aquaculture development. This helped streamline permit processing. He also emphasised that a systematic evaluation of management processes can be valuable in identifying institutional responsibilities at both government and private-sector levels, as well as for promotion of small-scale aquaculture.

Implications of Systems Approach for Future Aquaculture Development

Aquaculture can be socially, environmentally and economically sustainable and, contribute to the production of food and rural development, provided appropriate farming systems and management practices are adopted.

The multidisciplinary and multisectoral systems approach recognizes that technical, economic, social and environmental issues, as well as institutional factors, have to be considered in the process of development and management of aquaculture. The systems approach attempts to understand the way the system operates and the interactions between different components, and serve as a basis for better management. This multidisciplinary approach requires different skills and, as such, also needs cooperation among different disciplines and information exchange among different stakeholders. The application of a systems approach to aquaculture has several implications for a more analytical and structured approach to aquaculture development.

There is a need for better information on aquaculture systems and promotion of more effective information exchange between stakeholders. Information requirements include social and environmental interactions at the farm level, development of better practices targeting important environmental and social impacts, and seeking incentives for farmers to adopt better farming practices. Development of suitable indicators is also another important factor. For assessing natural resource use by aquaculture, recent research has emphasised the "ecological footprint" (Kautsky et al., 1997). The aims of the authors to develop a "sustainability" index are to be commended, however, there are a number of problems with the models promoted so far (Roth et. al., 2000), including:

* their static and dimensional (area) nature;
* the difficulties of incorporating economic and social values into the approach;
* the emphasis on biophysical factors;
* the lack of the footprint and ecological capacity to accept multisectoral or multiple uses;
* the lack of inclusion of ecosystem services (such as absorption capacity for discharge of nutrients into coastal ecosystems); and
* difficulties in making comparisons between different locations and systems.

There was a consensus in the Conference discussions on the need to develop and evaluate meaningful indicators of resource use efficiency to better define environmental interactions with aquaculture and improve management decisions. Research is required to develop and evaluate meaningful indicators of resource use efficiency useful to management decisions.

Institutional strengthening in the private and public sectors is another critical issue that must be addressed. The systems approach recognizes that national institutions (government and non governmental agencies) are the primary facilitators of aquaculture development in their countries. However, national institutions involved in the promotion of aquaculture need to increase their capacity and reinforce emphasis on multidisciplinary approaches. This will require a shift away from traditional "top down" and "technology focused" capacity building and extension approaches, towards an approach that is interactive and responsive to farmer needs.

As emphasised above, much aquaculture research to date has focused on technical questions, leaving the impression that the major constraint facing aquaculture is a shortage of technical knowledge. This detracts from the attention needed for development and basic aquaculture education (Edwards, 1998). Another important constraint to aquaculture is dissemination of existing knowledge, whether derived from research or the indigenous technical knowledge of farmers. The limited capacity of many developing countries' national institutions in education, research and development concerning the promotion of aquaculture compound this problem, and require serious attention in the new millennium.

Examples of the systems approach mentioned during the panel discussions included:

* its application to small-scale farmer research and extension;

* the development and application of best management practices to improve environmental management on commercial aquaculture farms;
* improving the profitability of aquaculture operations through the use of a structured, systematic approach to operations and business management; and
* the institutional use of a systems approach to assist in effective decision making at a regulatory/ planning level and within the aquaculture industry as a whole.

RECOMMENDATIONS OF THE BANGKOK CONFERENCE

The Following Recommendations were Adopted by the Millennium Conference

That a multidisciplinary and multisectorial and systematic approach be taken in the development of aquaculture and aquaculture research. Applying a systems approach allows the proper understanding and analysis of problems and opportunities, and the development of solutions based on the understanding of how systems operate. The systems approach involves basically seven steps:

* state the problem;
* identify the system;
* classify and describe the system, boundaries and key factors;
* analyse the problem;
* propose solutions;
* test the solution; and
* implement and disseminate knowledge to solve the problem.

Codes of Practice, and the Best Management Practices incorporated into such codes offer an opportunity to improve management. These need to be developed further and applied in real situations as a practical basis for the operating system. Special attention needs to be given to:

* implementation of Codes of Practice;
* the need for continual fine tuning to take account of new information and technology;
* a role for farmers in implementation (there should be greater participation by farmers and other end-users to provide feedback and fine-tune systems management practices); and
* the effectiveness of self-regulation.

There was a consensus on the need to develop and evaluate meaningful indicators of resource use efficiency to assist in understanding environmental interactions with aquaculture and improve management decisions. The use of the "ecological footprint" model as a tool was discussed, and it was suggested to develop a more appropriate index.

There should be more emphasis on multidisciplinary research using a systems type approach, to better understand problems and opportunities and develop more appropriate solutions to overcome aquaculture development constraints and unfulfilled potential.

There is a need to communicate practical examples where the systems approach has been used and to incorporate the problem identification and solving systems approach into educational and training programmes.

2

Aquaculture Products: Quality, Safety, Marketing and Trade

The growth of aquaculture has led to significant changes in how its products are perceived and marketed. In becoming an important contributor to the markets for seafood, aquaculture is increasingly subject to safety mechanisms and controls, such as the statutory hazard analysis critical control point (HACCP) methodology in certain developed regions. As both safety and trade regulations are harmonized at international levels, quantitative risk assessment and traceability will become integral components of aquaculture management. Developing countries have increased their share of the seafood export market to nearly 50 per cent of global trade, a significant portion being represented by aquaculture products (shrimp, salmon, molluscs, etc.), a percentage that should increase with the continued expansion of the sector. The long-term viability of aquaculture development will be market driven, accounting for consumer demand and the capacity to adapt to the structure and legislative demands of the target markets. Important externalities will affect the production sector in achieving its goals, including the topics of sustainability, traceability and interactions with the environment. The development of schemes of best practice that incorporate the addressing of such issues, alongside quality schemes and safety management, will aid the production sector to achieve its goals. Discriminatory tariffs on trade should be avoided, particularly given the increasing importance of aquaculture as an export-earner and contributor to food security. The sectoral response to such developments must include a better understanding of the complex issues faced by the production and marketing of consumer food products within highly competitive markets, where the assumption of responsibility is essential, both for the product and the actions taken to produce it.

Introduction

At a global level, aquaculture is one of the fastest growing food production sectors (9.6 per cent/yr in the last decade), a fact that will ultimately change the way that fish is perceived as food. A key element of this observation is the change in the supply opportunities for fish and fish products from a wild source to a cultured one. Aquaculture product expansion has placed increased requirements on quality and food safety by consumers and regulators. There are few questions about the evident nutritional benefits of consuming fish but, by the end of the 1980s, developed countries arrived at the conclusion that classic fish (and food) inspection procedures, based on the analysis of samples of the final product and on generic hygiene measures, were not enough to provide the necessary level of protection to consumers. A preventive system called HACCP (hazard analysis critical control point) was adopted, and governments started to shift their regulations to HACCP-based systems. If one word has to be chosen to explain HACCP, it is "prevention."

An important point within the global market is the growing importance of international agreements that involve food (and fish) safety aspects. About 40 per cent of all fish produced are traded internationally, which means that there is a search for common criteria that facilitate or permit clear rules for compliance.

The tendency is, therefore, to move towards the harmonization of national regulations, meaning that such regulations could assure an equivalent level of food protection to consumers. This is a relatively easy concept to understand, but very difficult to implement and validate in practice. In turn, this increases the importance of internationally accepted guidelines, recommendations and standards, such as those of the Codex Alimentarius. The provisions related to food trade of the General Agreement on Tariffs and Trade (GATT) Agreement compound this tendency, and all of these aspects are interlinked.

The extent of regional and international trade in aquaculture products is difficult to analyse because trade in many aquaculture products is not yet well documented in all of the main producing countries. Aquaculture contributes primarily to domestic consumption but, at an international level, important trade has developed for a number of aquaculture products, and this has been one of the principal driving forces for aquaculture development in many developing and developed countries.

International Agreements on Food Safety

While there is no common agreement on food or fish safety at the international level, agreements do exist on food trade that have implications for food safety and quality matters. The important one regarding food safety is the Agreement on the Application of Sanitary and Phytosanitary and Measures (SPS) (GATT, 1994), which introduces two very important concepts. The first is that "to harmonise sanitary and phytosanitary measures on as wide a basis as possible, Members (countries) shall base their sanitary or phytosanitary measures on international standards, guidelines and recommendations, where they exist... (Article 3)".

The same agreement establishes that, regarding food safety, "international standards, guidelines and recommendations" to be taken as reference will be those "established by the Codex Alimentarius Commission (CAC) relating to food additives, veterinary drug and pesticide residues, contaminants, methods of analysis and sampling and guidelines of hygiene practice".

The second important point of the SPS Agreement is the adoption of the criterion that risk assessment shall be the basis for determining the appropriate level of both sanitary and phytosanitary protection. In this case, the SPS Agreement imposes a strong condition on national regulations that have not yet accommodated this criterion. Current HACCP-based regulations include a "hazard analysis" step, but risk assessment, which should be a part, is performed only qualitatively, and regulations do not refer to risk assessment explicitly. It is clear that the future development and evolution of regulations will be in this direction. All developed countries and a large number of developing countries have taken up regulatory HACCP-systems. These concern the safety of fish and fish products, which include products from aquaculture, and it is assessed that approximately 65 per cent of the total international fish trade is performed under such regulations. The greatest exception concerns the Japanese market, which accounts for about 24 per cent of the total fish market (demand), but which has no HACCP regulations as yet.

FISH SAFETY IN THE EUROPEAN COMMUNITY

The Member States of the European Community (EC) are subject to both European and national legislation, and considerable developments in laws affecting aquaculture have occurred in the last decade. These are separated into two clear sections that concern the final product and the aquaculture process.

The HACCP-based regulation for fish and fish products in the EC introduced the concept of "own health checks" (HACCP in terms of the EC regulation relative to fish products), and it is the Council Directive 91/493/EEC that laid down the health conditions for the production and the placing on the market of fishery products. EC aquaculture production follows Council Directive 91/67/EEC, which concerns conditions of stock health and health management and which, in practice, generated a "family" of regulations. For instance, Council Directives 93/54/EEC, 95/22/EC and 95/70/EC have successively amended Directive 91/67/EEC. This "family" of regulations is aimed basically at controlling fish diseases; they are only, in a second instance, complementary with the HACCP-based regulations aimed at preventing human diseases. The regulations related to aquaculture production do not imply accomplishment of the fish safety regulations, however. Although there are no contradictions between either regulation "family", their complementarity and coordination will surely increase with time.

FISH SAFETY IN THE UNITED STATES OF AMERICA

The basic regulation in the United States of America, which makes mandatory the implementation of the HACCP system in fish and fish products, is that of the United States Food and Drug Administration (FDA,5).

This regulation does not mention fish obtained from aquaculture specifically, but this source is included implicitly in the definition of "fish" ("where such animal life is intended for consumption") and in other paragraphs of the regulation (e.g., when "drug residues" is listed as one of the possible hazards to be accounted for). However, the regulation is specifically centred on processing, and does not apply to aquaculture production in itself.

As a general conclusion, one can say that such regulations are structured around a core principle that defines the HACCP-based system, and that this basic principle is complemented by a number of regulations that can be defined as "regulatory standards". Each species and product may be defined within a "family" of regulatory texts that outline the minimum requirements for each particular case or procedure.

Fish and fish products from aquaculture are included, either explicitly or implicitly, in such HACCP-based regulations. However, whereas the HACCP system is well defined for processing steps or procedures once the fish has been harvested, the application of the HACCP system to the entire aquaculture production chain is not so clearly established within the regulatory environment. Developments in this area can be anticipated.

Further evolution of the fish safety regulations may be expected with the introduction of new technical, scientific and operative knowledge and, in particular, regarding mandatory quantitative risk assessment.

All matters that relate to safety, quality and trade rely on the ability to identify and trace the product, but the traceability of food remains as an enormous problem to be solved. In its broadest sense, a traceability system would provide and allow access to all information concerning a product. In the case of aquaculture, this could mean following the history of a production batch from the final point of sale back to the hatchery. This would mean assuring knowledge of the steps encountered in distribution, packaging, processing and harvest, including the growing conditions, the feeds and therapeutic agents used, the broodstock parentage and any other relevant information.

Clearly, such a huge amount of information cannot be encoded physically with the product, but needs to be accessible via coding and linking to appropriate databases. The ability to trace and isolate a problem with a product is essential to any safety system, but the concept of traceability is much more than this, since it affords assurance, generates trust and, very importantly, can allow the use of important information for marketing and sales. Information is a new measure for adding value and will increase in importance.

AQUACULTURE AND TRADE

International seafood exports reached US$48 billion in 1998 (provisional FAO data), up from US$36 billion in 1990, but down slightly from the figures for 1996 and 1997. The share of developing countries in seafood exports grew from 43 per cent to 49 per cent between 1990 and 1997, giving net receipts of foreign exchange that rose from US$10.2 billion to US$15.8 billion.

The rapid growth in aquaculture production has made the sector important to the economy of many developing countries, and it has become either an important source of supply, or the main supplier, in the case of some products. For these farmed products, production fluctuations have had a significant impact on price trends. In general, however, aquaculture products have helped to stabilize traded supplies and to bring down prices over the years. Furthermore, several species that were once considered to be high-value luxury products have now become more abundant through aquaculture production, lowering prices and expanding markets. The extent of regional and international trade in aquaculture products is difficult to analyse because trade in many aquaculture products is not yet well documented in all of the main producing countries. Furthermore, because international trade statistics do not distinguish between wild and farmed origin, the exact breakdown between fisheries and aquaculture in international trade is open to interpretation.

This situation will change gradually, as aquaculture associations emerge in the main producing countries and start to keep records, and in response to various trade regulations that distinguish between farmed and fished origin (e.g., for shrimp).

TRADED AQUACULTURE PRODUCTS

In 1998, the main internationally traded products from aquaculture were shrimp, salmon molluscs and seaweed. Other species showing strong growth are tilapia, seabass and seabream.

Crustaceans

Marine shrimp is the most prominent product from aquaculture in international trade, and aquaculture has been the major force behind increased shrimp trading during the past seven to eight years.

Shrimp is already the most traded seafood product internationally, with about 25 percent or 800 000 mt coming from aquaculture (FAO, 1999a). Since the late 1980s, farmed shrimp has tended to act as a stabilizing factor for the shrimp industry. Therefore, the major crop failures in Asia and Latin America during the past few years (caused by disease problems) have had a significant impact on overall supply, demand, prices and consumption trends.

The major markets are Japan, the United States of America and, to a lesser extent, the EC, while the largest exporters of farmed shrimp are Thailand, Ecuador,

Indonesia, India, Mexico, Bangladesh and Vietnam. Demand for shrimp is expected to increase in coming years, where Asian markets, such as China, the Republic of Korea, Thailand and Malaysia, will expand as local economies recover and consumers' demand more seafood. This trend is already reducing the availability of shrimp to traditional importers and will eventually put upward pressure on prices if supplies do not increase.

Trade in crab species has also increased with growing aquaculture production (1997: 165 000 mt). Especially important have been the exports of China (19,000 mt in 1998) to Hong Kong SAR China and Japan (INFOYU, 1999).

Finfish

In terms of total aquaculture output, finfish ranks first, with 49 per cent of the total production from aquaculture, of which the major part are carp species, which are consumed locally in the producing countries (mainly China and India). As opposed to shrimp, finfish aquaculture trade appears to be split between species having a high traditional demand and a "quality" image (e.g., salmon, seabass etc.) and convenience products (mainly fillets) of "cheaper" fish species (e.g., tilapia). The following species are the main products that are seen as being important in international trade:

Salmon: International trade in farmed salmon has increased from virtually nothing to more than 600 000 mt (1999) in less than a decade. The traded species are mainly Atlantic salmon and, to a much lesser extent, coho salmon, which accounted for 87 per cent and 12 per cent of 1997 salmon production, respectively (FAO, 1999b). The growth seen in trade has mirrored that of production, reflecting the fact that this activity is concentrated in a few countries that have limited domestic markets

(Norway, Chile and the United Kingdom). Norway, whose main market is the EC, is the main exporter of Atlantic salmon. On the other hand, Chile, whose principal markets are Japan and the United States, is the main exporter of coho salmon and the second largest exporter of Atlantic salmon (FAO, 1999b; FAO, 1999c; Kontali Analyse, 1999).

As production volumes have increased, so has competition within the market and costs and prices have been driven down. At the current price levels (+/-US$4/kg CIF [Cost, Insurance and Freight) Europe), salmon has become a relatively medium-priced product in international seafood markets.

Trout: Trout is traded internationally at a much lower level than salmon, with exports reaching 135 000 mt in 1998 out of a total production of 463 000 mt. (FAO, 1999b). However, trout production is split, approximately equally, between portion-size production and large trout, and where 70 per cent of global production is in Europe. Consumption, particularly of portion-size fish, is concentrated in trout-producing countries, but Norway, Chile and Finland have been able to farm particular qualities of large-sized, heavily pigmented trout for the Japanese market, primarily as a replacement product for coho salmon. Japanese trout imports reached 60 000 mt in 1998 (Japanese Marine Products Importers Association, 1999).

Tilapia: Tilapia species have also shown a tremendous growth in output (production of 946 000 mt in 1997), and although international trade is limited, it is growing. This is observed especially between Latin America (Costa Rica, Ecuador and Colombia) and the United States, as well as between Asian producers (Taiwan Province of China, Indonesia and Thailand) and the United States and Japan. There is also modest trade between the EC and producers in Jamaica, Israel and Zimbabwe. The biggest exporter, Taiwan Province of China, supplies Japan with high-quality tilapia fillets for the sashimi market and ships frozen tilapia to the American market. Taiwan Province of China now exports about 35 per cent of its domestic tilapia production and supplies 80 per cent of the United States tilapia imports (1998). Thailand and Indonesia export less than five per cent of their production (Dey and Eknath, 1997). Vietnam has also recently entered the world tilapia market, and China exported 500 mt to the United States in 1998 (NMFS, 1999).

Tilapia has become the third highest imported aquaculture product, by weight, in the United States (1998 imports of 28 000 mt), after shrimp and salmon. United States imports rose 14 per cent by quantity in 1998 (NMFS, 1999). In the long term, tilapia prices are expected to decrease, and this should lead to greater exports to the United States, as well as to Europe, which is seen presently as being undeveloped as a market for tilapia.

Seabass and Seabream: In Europe, the marine fish farm industry of the Mediterranean has focused on the production of European seabass and gilthead seabream, and it intends to copy the success of salmon growers. Production reached almost 90,000 mt in 1999 (FEAP, 1999), of which nearly 90 per cent was exported from the country of origin, mainly to Italy and Spain. The principal exporter was Greece, which exported about 70 per cent of domestic production. Italy was, for a long time, almost the exclusive market for Greek production. However, as a result of market development efforts, some 15 per cent of Greek exports are now going to "new" markets (e.g., the United Kingdom, Germany, France etc.), and the share of these markets is expected to grow. It is not to be ignored that trade in live fingerlings is made from hatcheries in Italy, Spain and France to farms in Greece, Malta, Croatia and elsewhere.

As output of seabass and seabream has grown, market prices have more than halved between 1990 and 1999, dropping from US$16/kg to less than US$5/kg today. The rapid saturation of the market and the speed of the parallel decline in prices (down 50 per cent in five years, compared to the ten years for a similar drop in the case of Atlantic salmon) is attributed to several factors:

* the much smaller traditional market for these species (southern Europe), compared to Atlantic salmon;
* lack of diversified products;
* inadequate marketing and market development; and
* absence of technological advances (e.g., genetic improvement, efficient feeds and feeding strategies etc.) that could significantly improve productivity.

The substantial drop in prices should help to open new markets and expand existing ones, provided that acceptable profit margins can be sustained in production through improvements in productivity, diversification of products and intensified marketing efforts. American Catfish: American catfish is now the fifth most consumed fish species in the United States, measured at 0.4 kg/capita edible weight in 1998 (NMFS, 1999). Exports are limited, as production targets the domestic market, but some exports have started to Europe. The reason for the success of catfish is similar to that of tilapia: consumer demand for white, easy-to-prepare fillets.

Seaweed

Farmed seaweed production has increased in the last decade (6.8 million mt in 1997) and is now 87 per cent of total seaweed supplies. Most of the output is used domestically for food, but there is growing international trade. China, the major producer, has started exporting seaweed as food to the Republic of Korea and Japan. The Republic of Korea, in turn, exports some quantities of Porphyra (red seaweed) and Undaria (brown seaweed) to Japan (1998 exports of 26 000 mt). Chile produces over 100 000 mt, and is responsible for some 75 per cent of global production of Gracilaria sp., which is used for agar manufacture. Significant quantities of Eucheuma (red seaweed) are exported by the Philippines, Tanzania and Indonesia to the United States, Denmark and Japan. Total EC imports of seaweed in 1998 amounted to 58,000 mt, with the Philippines, Chile, Indonesia and Australia as the biggest suppliers.

Molluscs

International trade in molluscs is limited when compared to total output, with less than 10 per cent of production traded. The major importing markets are Japan, the United States and France, while major exporters are China and the Republic of Korea. The contribution of these farmed products to trade is uncertain.

Oysters, followed by clams, scallops and mussels, lead farmed mollusc production, but international mollusc trade concentrates on scallops and clams (fresh and frozen). Total fresh and frozen scallop imports have grown from 28 000 mt in 1985 to 65 000 mt in 1998, reaching US$510 million. Clam imports have risen from 33 000 mt to 152 000 mt in 1997, valued at US$257 million. Mussel imports have shown a downward trend after a peak of 175 000 mt in 1992, but have bounced back to 188 000 mt worth US$220 million. Oyster imports have grown from below 10 000 mt in 1985 to 28 000 mt (US$112 million) in 1998.

Live Seafood

The cultural preferences and growing affluence in Asia indicate a clearly positive long-term trend of live seafood commerce. The live seafood market is largely restricted to the restaurant trade and to consumers with a relatively high disposable income. Major market expansion is anticipated due to demand in China, but is also expected in Malaysia, Singapore and Taiwan Province of China, as well as in parts of North America with large Chinese communities. The potential for aquaculture to supply the market is promising. The sector is already supplying large amounts of shellfish and limited quantities of grouper, crabs and other species. Technological developments in the culture of preferred live food species will increase the contribution of aquaculture to supplies (Riepen, 1997).

Seed Supplies

There appears to be significant regional and international trade in the seed of cultured aquatic organisms, mainly from aquaculture sources, but this is poorly documented at present in most instances. Mention has been made above of the regional trade of fingerlings for Mediterranean seabass and seabream, but there is also international trade in wild glass eels (e.g., the recent development of large purchases of European eel elvers by China), "eyed" (fertilized) eggs of salmon and trout, the postlarvae of various cultured shrimps, Indian and Chinese carps, and others. There is also limited trade (in terms of quantity) in broodstock. Documentation of trade in seed and fingerlings will improve gradually as a response to concerns about the spread of pathogens and the movement of genetic material. One would anticipate that the concerns of traceability will also support such documentation.

ISSUES AFFECTING FUTURE TRADE IN AQUACULTURE PRODUCTS

The long-term viability and sustainability of aquaculture development, particularly in respect of commercial aquaculture, will be market driven, taking into account not only the consumer's requirements but also the structure and legislative demands of the target market, be it local or international. Some of the key issues that need to be addressed are mentioned in the following sections.

Externalities

Environmental and social concerns can influence markets for consumer goods and have already influenced farmed shrimp exports to North America and Europe in recent years. There is a growing desire for knowledge of what is being consumed, a position that, in some cases, is accompanied by accountability for consumption.

The importance of attaining sustainable aquaculture with no or limited negative externalities will force many exporting countries to adopt more sustainable production practices. The introduction of eco-labelling schemes will further accentuate this trend. Where aquaculture is perceived as a non-traditional food-producing sector, it will have to further establish its credentials for sustainability, particularly when compared to fisheries and crop agriculture. This consideration requires to be extended to safety assessments, based on risk assessment and the precautionary approach, before entry into production of new or exotic species, including the potential use of the products from modern biotechnology.

Awareness of environmental and welfare issues is increasing, particularly in the developed countries, where purchase decisions can be influenced by adverse publicity or a lack of information. As livestock farmers, aquaculture producers are increasingly required to act responsibly and in line with standards expected of the activity. While such topics are partially addressed by the Food and Agriculture Organization's Code of Conduct for Responsible Fisheries (CCRF), it is increasingly recognized that standards applicable to international trade in aquaculture require harmonization. At national levels, safety and quality management systems including Good Agricultural Practice (GAP) and Good Manufacturing Practice (GMP) should be developed and put in place in order to assure the production, distribution and sale of aquaculture products that are safe to consume and of high quality. Such measures require active and competent professional associations, working hand in hand with government, in order to be successful.

The collection, analysis and dissemination of relevant information should be facilitated in order to enable producers and industry operators to make informed decisions and to ensure consumer confidence in aquaculture products.

Quality and Safety

With growing concern about food safety, increasing efforts have been undertaken to improve the quality of food that is placed in the market, which evidently includes aquaculture. International codex standards cover aquaculture products, and the introduction of mandatory HACCP requirements for exports to the United States of America and the EC in 1997 has already had a great impact on trade in several aquaculture products.

Some countries have developed comprehensive HACCP plans for selected aquaculture products; for example, the United States now has plans for catfish, crayfish and molluscs. In other countries, individual aquaculture producers have undertaken voluntary certification

(ISO 9000) for control as well as marketing purposes. Such certification appears to be increasingly required for entry into markets such as multiple retail stores.

Alternative efforts include the development of industry-led quality schemes, which require government approval, to which individual producers can adhere. These schemes have controllers and strict operating procedures and conditions, serving to provide products of high quality and known origin.

In the field of HACCP, focus is increasingly on risk assessment by the operator, and this issue will put further institutional demands on exporting countries.

The effects of bovine spongiform encephalitis (BSE-mad cow disease) in cattle, the debate over genetically modified organisms (GMOs) and the increased awareness of toxin presence in the environment and food have contributed to a higher consciousness of the consumer to the quality and content of food in general, especially in the developed countries. Actions to assure best practice, including traceability throughout the entire supply chain, are seen as being inevitable for assuring both the credibility and the sustainability of the aquaculture sector. This should not just apply to the aquaculture production process, but also include the feed content, particularly where additives and the potential use of GMO components are concerned.

3

Sales and Marketing

Attaining and maintaining consumer confidence requires considerable effort. It is no longer satisfactory to believe that production of a "high value" aquaculture product is a guarantee for the producer's long-term success.

Aquaculture history shows how technical success for rearing a "high-value" product incites production expansion, which leads to price drops or crashes. There are many different opinions as to the "whys" and "wherefores" of these circumstances, and market saturation is often given as the answer. There are, however, other explanations.

The long time required between investment in stocks and financial returns from sales can force producers to sell too early or at low prices, simply because of current cash needs, a circumstance that is punctual and rarely accompanied by marketing efforts. Such situations lead to market surpluses and price crashes in circumstances where better financial and production planning could assist.

The geographic dispersion of the activity and the relatively small amount of production per unit mean that there are a high number of sellers to markets that have few buyers (in number), a situation that gives keen competition and an advantage to the buyer. In addition, the aquaculture sector is relatively young and spends little money on marketing and promotion, particularly at the producer end of the scale.

Aquaculture produces perishable products with a short shelflife, particularly where products are sold fresh, meaning that distribution skills and production planning (to avoid surplus supplies to markets at specific times of the year) have to be honed to meet market demands.

The need for improved marketing and distribution skills is evident in many individual cases, where the grouping of producers under cooperative structures or as producer organizations is seen as being a response to such criticism.

The creation of efficient marketing systems, in which prices and costs are determined by supply and demand, moving to assure economic efficiency and sustainability, should be facilitated. The use of economic incentives and disincentives should be used to rectify market failure and the unsustainable use of resources.

Distribution Channels

The growing market share of multiple retail stores (super-and hypermarkets) in the distribution of foodstuffs has significantly changed patterns of production, supply and distribution.

For fish and fish products, these changes have had, in many markets, a profound impact on both the demand for products from aquaculture and the production sector itself. Modern distribution channels have developed buying criteria with precise requirements for quality, portions and sizes, price and delivery times that often can only be met by aquaculture producers. This explains, to some extent, the success of products such as Atlantic salmon, which is prevalently sold through super-and hypermarkets in European markets. This has led to the virtual disappearance of the specialized fishmonger in certain developed countries and has imposed significant changes on the profession, in operating, marketing and organizational skills.

An example of this is the demand for packaged products, even for fresh, gutted fish. The demand now focuses on fixed weight packs (so that no individual weighing is required), particularly for processed products, using modified atmosphere packing (to assure freshness and a longer shelf-life) and where the sale price is attached prior to despatch to the supermarket. These conditions mean that, for the aquaculture producer to be capable of attaining this market, access to approved processing and packing facilities has become an essential part of working, as opposed to an option.

Labelling should follow the recommendations and codes of practice that are in line with the international requirements of the World Trade Organisation (WTO) and the Codex Alimentarius, demonstrating full traceability.

Tariffs

Despite steady reductions in customs tariffs on both fish and aquaculture products in recent years, tariffs and import licenses continue to represent barriers to trade in many countries. This is especially the case in many fast-growing economies in Asia, but important markets such as Japan, the European Union and the United States of America all give competitive advantages to domestic producers of many species, especially in the case of processed products. On the other hand, producers in the developed markets argue that they are subject to higher levels of control and imposed regulatory investments that increase their production costs.

Average tariffs on imports from developing countries are now estimated at 4.8 per cent, a cut of 27 per cent from previous levels (FAO, 1995). The long-term trend, with growing membership in the WTO, will lead to further tariff reductions. The multilateral trade negotiations, which were to have started in late 1999, now show some delay in taking off, but could also have a large impact on future trade in fish and aquaculture products.

Measures that are taken to protect human life, animal life and health, the environment and the interests of consumers are increasingly seen as being potentially discriminatory, either in the nation of origin or in the exporting nation, depending on the point of view and the legislation in question.

Food Security

Aquaculture is an important source of seafood, and the major part of the total output from aquaculture is consumed internally by the nations that produce, providing employment and an important nutritional contribution to society. Aquaculture has also become a significant source of foreign currency for many developing nations, since the products exported are usually the more valuable ones destined for markets in the developed world. These revenues allow the countries to import other less costly protein and, as such, aquaculture plays an important role in food security, even where significant proportions of the output are exported.

Conclusions

The circumstances of producing and marketing fish and seafood produced from aquaculture are changing quickly. Technological advances have brought new species and higher productivity to the sector, which has developed to become an important contributor to national and international markets. Consumer demand for specific products, combined with good business opportunities, has contributed to the rapid development and restructuring of certain aquaculture subsectors, notably those concerned with export to developed countries. The industrialization of production and processing, alongside increased consumer awareness and legislative actions, has imposed quality and safety standards that are becoming requisite rather than optional. Sustainability and environmental friendliness are also factors that are being linked to sectoral acceptability and, hence, influence quality, marketing and trade issues.

The complex nature of the activity and its diversity mean that increased interregional and international cooperation should be encouraged, particularly in the fields of safety, quality and trade, where accent should be given to promoting a harmonized sectoral approach to the issues of concern.

4

Current Status and Development Trends

INTRODUCTION

The past 20 years have seen Asian aquaculture evolve from a traditional practice to a science-based activity and grow into a significant food production sector, contributing more to national economies and providing better livelihoods for rural and farming families. During the period 1988-1997, Asian aquaculture production grew from 13.4 to 32.6 million mt (APR [Annual Percent Rate of growth] 11 percent), increasing its value from US$19.3 billion to US$42 billion (APR 9 percent).

Aquaculture used to be regarded as an infant when compared with crop and livestock farming and capture fisheries. In most parts of Asia, the activity has matured to become a clearly defined economic sector that is better organized and characterized by state patronage and strong private-sector participation. Many changes have occurred alongside the sector's development. Aspirations for higher yields through technological innovation have been tempered with concerns for sustainability. Economically, the drive for higher profits has been qualified by schemes to distribute benefits fairly. As a commodity industry, the purposes of producing more food, earning higher incomes and improving economies have been expanded to include ensuring food security, alleviating poverty, and promoting social harmony and prosperity.

These shifts in outlook, not confined to aquaculture and prompted by global and social forces, occurred in the last part of the previous century. It is in the midst of these changes that aquaculture development in Asia finds itself at the beginning of the new millennium.

Status of Asian Aquaculture

Asia dominates global aquaculture production. This simple statement covers, however, the extreme diversity encountered within the sector, not only in the species reared and the technologies and farming systems employed, but also in the role and objectives of aquaculture development in different countries, the priorities placed by governments and the threats and constraints to its growth.

These considerations, in addition to the wide range of geographic locations, habitats and levels of national development, are important influences on how, and how fast, aquaculture will develop in the region. For these reasons, it would be better to look at the status and trends of Asian aquaculture by subregion, or on a national basis, with China classified by itself as a subregion in view of the sheer size of its aquaculture.

By the same token, India merits the same focus. Nevertheless, analysis of the overall regional status serves a useful purpose, particularly in trying to identify regional opportunities and actions.

Asia's contribution to total world production in 1997 (Fig. 1), following the International Standard Statistical Classification for Aquatic Animals And Plants (ISSCAAP) grouping, was: finfish-89 percent; crustaceans (marine)-80 percent; freshwater crustaceans-94 percent; molluscs-88 percent; aquatic plants-98 percent, and miscellaneous animals and products-99 percent.

The region provides 91 percent of global aquaculture production, where the top 10 Asian aquaculture producers are China, India, Japan, Republic of Korea, the Philippines, Indonesia, Thailand (also the top seven producers in the world), Bangladesh, Viet Nam and DPR Korea.

In 1997, the combined aquaculture production in Asia was 32.63 million mt (Fig. 2) valued at US$41.95 billion, an increase of 144 percent and 117 percent in weight and value, respectively compared to 1988. Aquaculture production in the region has been growing at a rate nearly five times faster than landings from capture fisheries, and its share of total fisheries landings in the region increased from 32 to 50 percent between 1988 and 1997.

PRODUCTION BY COMMODITY GROUPS

Finfish

Throughout Asia, the production of finfish has increased from 6.36 million mt in 1988 (having a value of US$10.1 billion) to 16.7 million mt (US$22 billion) in 1997, representing an average APR of 11 percent, while annual growth during the period 1992-1996 averaged 15 percent. In South Asia4, fish for food, mainly freshwater species, dominate production, with over 94 percent of total fish production in 1997. Freshwater and diadromous species occupy 92 percent of the subregion's total production, the remainder being crustaceans. Subregional production attained 2.2 million mt in 1997, worth US$22 billion.

In Southeast Asia, freshwater foodfish species comprise almost 30 percent of the total aquaculture output, mainly from Indonesia, Viet Nam, Thailand and the Philippines. While the overall production of diadromous species has been decreasing, milkfish contributed an additional 13 percent in 1995. Rising Indonesian aquaculture has been a major contributor to the increased production of freshwater species in this subregion. Aquaculture produced 1.5 million mt worth US$3.4 billion in 1997.

China is the major finfish producer in East Asia6, where total production was 13 million mt in 1997 (Fig. 3) worth US$16.1 billion. The production of freshwater finfish in China, mainly of carp and tilapia species, was over 12 million mt in 1997, accounting for 37 percent of regional aquaculture and 34 percent of global aquaculture (Fig. 3). In recent years, diversification towards luxury freshwater species has been emphasized, including mandarin fish, freshwater crabs and prawns, eels and softshelled turtles. Nonetheless, carps accounted for 92 percent of freshwater production and represented 47 percent of national production in tonnage and 40 percent in value.

Japan produces some 322 000 mt of finfish, worth US$2.8 billion, having an average value regularly higher than US$9/kg. This is mainly due to the production of significant quantities of Japanese amberjack (138 000 mt), silver seabream (80 000 mt), Japanese eel (24 000 mt) and rainbow trout (13 500 mt).

Around 14 percent of all finfish production in the world consists of carp, more than 90 percent of which is produced by aquaculture and where 80 percent of this amount is produced by China. In most countries, carp is consumed locally and, with few exceptions, producers of

Chinese and major carp species have been unable to find markets outside Asia. Carps, as a group, are not traded globally. Carp supply in China, India and countries of the CIS [Commonwealth of Independent States], will probably continue to increase steadily, at least in the near future, in response to population growth and demand.

Tilapia has spread outside of Africa, its origin, and is now a common feature of aquaculture not only in Asia but also in Latin America and the Caribbean. Producers have access to domestic markets in several Asian countries, such as the Philippines, Thailand and China, and there is also an established and rapidly expanding market in the United States that is providing additional impetus to tilapia farming in several countries.

Small quantities of trout are farmed in Iran, Pakistan, the colder regions of India, and in Nepal, but Japan is the largest producer with some 13 500 mt produced in 1997. Trials are starting in Thailand to grow trout in the cooler climates (northern hills) as an income generating activity and largely for the domestic market.

The production of eels is significant in China, Japan and Taiwan Province of China, where the focal point for trade is Japan, which imported 11 000 mt of live eels and 45 500 mt of processed eels in 1996. Regional production has risen from 92 000 mt (1988) to 224 500 mt (1997). However, major growth has only occurred in China (>100 000 mt over the period), production dropping significantly in Japan (from 40 000 mt to 24 000 mt) and in Taiwan Province of China (52 000 mt to 22 000 mt). Furthermore, the 1998 value of the eel output (US$314 million) in Japan was a huge drop from the 1995 peak of US$577 million, because of the decrease in quantity produced (from 29 131 mt in 1995 to 21 971 mt in 1998) and average price of the product from (US$20/kg to less than US$15/kg.

As production capacity has grown, disease and a shortage of elvers of Japanese eel (Anguilla japonica) have caused seed supply problems in the region and initiated the trade of elvers of the European eel (A. anguilla) from Europe to Asia, particularly, China. Recently, a resurgence in the recruitment of the Japanese elver has alleviated the situation.

Crustaceans

Aquaculture of crustaceans, mainly shrimp and crabs, yielded 539 000 mt in 1988, worth US$3.5 billion. By 1997, this had risen to 1.1 million mt (APR 9 percent), having a value of US$7.2 billion (APR 9 percent).

Shrimp production in 1997 was worth US$6.9 billion at farm gate prices to the region, representing a total of 902 000 mt. The value added through processing is also very significant, making it worth more than US$6.9 billion to the region. Penaeid shrimp is the main crustacean aquaculture product that is traded internationally but, although this is an important source of hard currency in many developing economies, shrimp are much less significant as a food source for domestic consumption. The total volume of cultured penaeid shrimp is now close to half that produced by capture fisheries. In volume terms, the rate of increase in production of penaeid shrimp is tapering off in Asia, but one can see a significant increase in the production of freshwater crustacean species, Macrobrachium rosenbergii, mainly in Bangladesh and China.

The growth of penaeid shrimp production has been predicted to slow down in the immediate future. One of the main reasons for this short-term prediction is the slower rate of economic growth in Japan, the world's largest market for shrimp, as well as the slower economic growth expected in the other developed economies. In addition, the management of shrimp culture is not at a uniformly high level. Furthermore, the emergence of stringent environmental regulations may slow down expansion in Asia.

In South Asia, crustacean culture is dominated by penaeid shrimp, which increased from 37 000 mt in 1988 to 134 000 mt in 1997 as a result of increased production in India, Bangladesh and Sri Lanka valued at US$858 million. Growth slowed in 1997 due to disease outbreaks and India's sanctions on shrimp culture. Of note is the large harvest reported in 1997 for the freshwater giant river prawn (M. rosenbergii) in Bangladesh, 48 000 mt valued at US$497 million.

In Southeast Asia, the production of shrimp and prawns grew from 220 000 mt in 1988 to 562 500 mt in 1997, where Penaeus monodon (giant tiger shrimp) represented 450 000 mt in 1997 (80 percent). The decade reviewed (1988-1997) shows the phenomenal growth of this sector in Thailand (+242 percent), the world's leading producer of this species, Viet Nam (+293 percent) and Indonesia (+115 percent). After production stagnation during the late 1990s, Thailand has rebounded with production of an estimated 250 000 mt in 2000.

East Asia has witnessed significant reductions in shrimp production, where production of the fleshy prawn (Penaeus chinensis) dropped from a high of 207 000 mt in 1992 to a low of 64 000 mt in 1994, recovering to 146 000 mt in 1997. Taiwan Province of China's giant tiger shrimp production dropped from a peak of 78 500 mt in 1987 to only 5 000 mt in 1997. Overall, the subregion's production of shrimp and prawns has decreased from 246 000 mt to 163 000 mt, but has shown signs of recovery in recent years. Other crustaceans for which production is increasing are crabs, notably in China, and after a period of decline, production of freshwater prawn is also rising with increasing outputs from China and Bangladesh. In West Asia, crustacean production is nominal and restricted to penaeid shrimp. In Iran it is small but increasing.

Molluscs

Mollusc aquaculture is dominated by East Asia (98 percent of the region's production), the remainder being attributed to Southeast Asia. Regional production reached 7.6 million mt in 1997, valued at US$7.6 billion, compared to the 1988 production of 2.6 million mt worth US$2.5 billion, an APR of 13 percent in volume and 14 percent in value.

Molluscs are almost always reared for sale, rather than household needs, usually being sold to markets located near the place of production. Mollusc culture in Asia was not affected much by the 1997-1999 economic downturn and will continue as a source of growth in aquaculture production. Cockles and mussels are promoted as cheap sources of protein, while oysters tend to be more of a luxury item, consumed domestically and in some countries as an export item Pearl oysters are significant industries in China, Indonesia, Japan and the Philippines and to a lesser extent in India and Thailand.

Big increases in production have occurred throughout the region during the period examined, but most significantly for:

* Japanese carpet shell (China: from 186 000 mt to 1 257 500 mt);
* razor clams nei (not elsewhere included) (China: from 138 500 mt to 354 000 mt);
* Pacific cupped oyster (China: from 447 500 mt to 2 328 500 mt); and

* Yesso scallop (China: from 129 500 mt to 1 001 500 mt; Japan: from 180 000 mt to 254 000 mt)

Elsewhere in the region, oyster production has stagnated. Mussel production, in general, has declined everywhere, with the exception of the increases reported for the Korean mussel in the Republic of Korea (8 000 mt to 64 000 mt).

Aquatic Plants

East Asia and Southeast Asia produce the bulk of aquatic plants within the region, where seaweeds, mostly brown or red, dominate the volumes reported.

These subregions provide almost all of the brown, red and green seaweeds produced in global aquaculture. While aquatic plant production almost doubled between 1988 and 1997 in East Asia, rising from 3.5 to 6.3 million mt, the contribution to aquaculture production dropped from 32 percent to 23 percent. In Southeast Asia, the pattern is different, production growing over the same period from 0.34 to 0.76 million mt, the contribution changing from 22 percent to 25 percent. The region's total production was 7 million mt in 1997, valued at US$4.8 billion, figures that indicate APRs of 8 percent and 5 percent, respectively. The average value/kg for this group has eroded from US$0.84 to 0.68, an APR of-0.2 percent.

The highest reported global production of any cultured aquatic organism for 1997 was the kelp Laminaria japonica, totalling over 4.4 million mt. In the top 10 species, carps have consistently led production, followed by plants and shellfish (1988); by 1997, shellfish had overtaken aquatic plants in this position.

Comparison of the top ten aquaculture species in the Asian Region, ranked by production importance in 1988 and 1997 is given in Table 1.

One can note that the top ten species contribute two-thirds of the region's production, but that carps have displaced aquatic plants from its dominant position and shellfish production, having tripled in volume, now occupies second place in the groups containing the top species.

Most notable is the rise to the top, from 5th place, of the giant tiger shrimp (Penaeus monodon), which contributed 9 percent of the total value of Asian aquaculture in 1997, although it is 17th in the production ranking. Thailand contributed more than half this figure, despite disease problems. The production of P. chinensis, which is predominantly cultured in China, was ranked 42nd.

While the production and contribution of the major carp species have increased significantly, the average US$ value/kg has gone down; for the examples cited in this analysis, reductions of 12-30 percent have been seen.

Most of the shrimp species have maintained or increased their US$ value/kg. During the same period, the value of the Japanese eel has halved, while certain fish species (milkfish, rohu, Japanese amberjack and tilapias) have increased.

Environment

If one separates aquaculture production by environment, the proportion of freshwater species in total Asian production increased from 43 percent to 49 percent between 1988 and 1997, an average APR of 12 percent. Brackishwater production had an average yearly growth of 5 percent and marine production 10 percent.

The proportion of brackishwater aquaculture decreased from 7 percent to 4 percent while experiencing growth at an APR of 10 percent during the same period.

Similarly, marine production had average growth of 10 percent, but its share decreased from 50 percent to 47 percent.

Freshwater finfish, in particular the Chinese and major carps, accounted for the greatest share (48 percent) of regional aquaculture production in 1997, while aquatic plants contributed 22 percent. A key factor in the rapid rise in the production of some of the species cited has been the increased availability of hatchery-produced seed.

Main Culture Systems

Inland fresh-and brackishwater ponds, tanks and reservoirs comprise the largest production areas in Asia, with some 1.99 million ha in China. Similarly, there are 147 000 ha in Bangladesh, 900 000 ha in India, 462 000 ha in Indonesia, 255 000 ha in the Philippines, 320 000 ha in Thailand and 392 000 ha in Viet Nam. Viet Nam, Indonesia and the Philippines have more brackishwater than freshwater ponds and which are devoted to shrimp culture as well as other species, such as milkfish in Indonesia and the Philippines.

Freshwater ponds in Viet Nam, Cambodia, Laos, Bangladesh and India are generally small and used to grow food fish that are low on the trophic level. These are usually integrated with other operating systems, such as the 127 000 ha of VACs (vegetable-fish-livestock plots or gardens) in Viet Nam, and the undrainable composite-culture ponds in India.

The Network of Aquaculture Centres in Asia-Pacific/ Asian Development Bank Study (ADB/NACA, 1998) found that almost all carp farming in the region is a polyculture of indigenous and (often) exotic carp species, with the exceptions of Korea and Indonesia, where monoculture of common carp was common. The carp farming systems also commonly contained other non-carp species.

There are three distinct areas of carp polyculture in Asia: South Asia has the major carps; China and cooler waters use Chinese carps, and Southeast Asia uses a mixture of exotic and indigenous carps. For example, in Cambodia, Laos and Thailand, indigenous cyprinids such as Puntius spp. are more common. Thailand's freshwater ponds are operated for a mixture of objectives. The products of school and village ponds provide food and income among poor families, and these comprise nearly 13 000 units, ranging in size from one-third of a hectare (school ponds) to 8 ha (community ponds). All in all, Thailand has a total of 247 300 ha of inland aquaculture ponds, comprising 122 000 farms and producing annually 196 000 mt of mostly food fish. Commercial freshwater ponds are used to raise higher-priced indigenous species, such as catfish, snakehead, freshwater prawn and tilapia. Many of the commercial entities are integrated farms.

Cambodia has the highest percentage (97 percent) of integrated farms within the semi-intensive carp production activity, perhaps a result of the introduction of this system by nongovernmental organizations (NGOs) as opposed to tradition. Thailand follows with 62 percent of farms integrated, primarily with chicken farming and tree crops.

Brackishwater ponds are generally devoted to higher-value species, shrimp being the most popular, followed by other crustaceans and diadromous species. Indonesia has extensive brackishwater ponds (385 000 ha] used for shrimp and milkfish. The Philippines has more than 250 000 ha producing milkfish and shrimp; lately, trials have been made with salt-tolerant tilapia being rotated or polycultured with shrimp to determine its effects in controlling shrimp diseases. In India and Bangladesh, there is a tradition of extensive culture of shrimp in large water bodies.

The potential for integration of rice and aquaculture has long been recognized. China has shown that rice-fish systems are indeed productive and, by also using the system to nurture fingerlings, can generate additional income for rice farmers over a short period. China has 1.3 million ha applied to the paddy cum fish system and an additional 281 000 ha of paddy used for fingerling growing. China's production from the rice-fish system was 455 000 mt in 1997 or almost 500 kg fish/ha.

Freshwater fish culture integrated with rice is also found in South Asia and Southeast Asia. Alternate cropping of rice and shrimp-a crop of rice in the wet season followed by a crop of shrimp in the dry season-is also found in some seasonally brackishwater deltas in Southeast Asia (notably Viet Nam) and South Asia (India and Bangladesh).

China's freshwater fishponds yield an average of nearly 5 mt/ha (9 million mt of fish from 1.9 million ha). Culture-based fisheries in open waters are likewise very productive; 880 000 ha of freshwater lakes yielded around 810 000 mt of fish in 1997, and 1.56 million ha of reservoirs provided 1.16 million mt. China has also exploited the potential of canals, producing 602 000 mt of fish from 370 000 ha of these waterways. The system used within canals usually involves cages or pens.

The NACA/ADB study on aquaculture sustainability and the environment (NACA/ADB, 1998) found that only 24 percent of surveyed freshwater farms in China were integrated, reflecting farm intensification and perhaps, changes in traditional farming practices. The report cautions that this observation needs further confirmation. China's mariculture (in shallow coastal waters and bays) is likewise characterized by the integration of various species (e.g. seaweed, mollusc an cucumber within a column of coastal water) for an intensive use of the marine environment.

In 1997, China harvested almost 4.7 million mt of seaweeds, 6.5 million mt of molluscs (including scallops and abalones) and 21 500 mt of other coastal aquaculture species. The areas where mariculture is done were made up of 7.9 million ha, including shallow coasts (3.7 million ha), bays (0.5 million ha) and mudflats (3.7 million ha).

Culture-based fisheries in open-water systems, such as oxbow lakes, are significant in India and Bangladesh, where production yields are around 500 kg/ha for native and exotic carps. Stocking involves both native and exotic carps, although some recent stocking programmes in Bangladesh are giving more emphasis to indigenous carps due to concerns over the potential impacts of stocking of exotics on native fisheries and aquatic biodiversity.

Production Losses

The first attempt, at a regional level, to quantify the effects of disease occurrence was made by NACA (ADB/NACA, 1991) in 1990. At that time, losses incurred from disease outbreaks were estimated to have been worth US$1.4 billion a year in the developing countries of Asia, and these have increased since. Reports from China suggested that in 1993 alone, shrimp diseases caused losses of US$1 billion. An ADB/NACA farm survey in 1994-1995 (ADB/NACA, 1998) indicated that for carp and shrimp farming, losses from diseases and environment-related problems were valued at around US$300 million a year. Furthermore, external influences that are associated with disease problems (including chemical use, environmental impacts, trade issues etc.) have yet to be properly assessed. Generally, economic losses are likely to increase in scale as aquaculture expands and intensifies, affecting both large-scale and small-scale production units.

Production Trends and Growth Rates

A noteworthy, but perhaps disturbing trend seen in the statistics for the Asian countries/territories from 1988 to 1997 is that the average aquaculture growth (APR) has been 10.4 percent by weight and 9.0 percent by value, but production has not shown steady growth, with year on year increases varying between 1.6 percent in 1990 and 18.6 percent in 1992. After peaking at 18-19 percent in 1992-1993, the rate of increase has slowed down steadily to 5.7 percent in 1997.

This observation coincides primarily with the trends seen for Chinese aquaculture. Some countries within the region have shown production decreases, with a steady decline of production through the period. These are Japan, China, Hong Kong8, Democratic Republic of Korea and Taiwan Province of China.

A striking trend in Japanese aquaculture has been the decline in production of the major species. For instance, the yields of Japanese amberjack have declined from a high of nearly 170 000 mt in 1995 to 147 000 mt in 1998, while laver dropped from a peak of 483 000 mt (1994) to under 400 000 mt (1998). Pacific cupped oyster production has also shown a steady decline. Of the major mariculture species, only silver seabream and Yesso scallop have shown trends towards rising production.

The annual rate of production growth has been consistently higher in China than in the rest of the region, reflected by the increased rate of difference in the compared production volumes. If aquaculture values are compared, the picture is slightly different. This figure shows that the difference in value is proportionately much less than that of production, confirming the effects of diversification and the maintenance of value in aquaculture in the rest of Asia. Indeed, in China the APR for value has grown and shown stability from 1990-1996, dropping off in 1997. In the rest of Asia, the value APR grew steadily from 1989, but has since declined regularly from 1994 onwards, even showing negative figures in 1996 and 1997. Theses circumstances meant that Chinese production alone had a total value higher than the rest of Asia from 1996.

Demand for Aquatic Produce

The projections for the average annual population growth rates in the region from 1998 to 2020 are:

* 0.8 percent for East Asia, growing to 1.66 billion;
* 1.6 percent for Southeast Asia, growing to 655 million;
* 1.7 percent for South Asia, increasing to 1.72 billion; and
* Iran will have 89 million and Australia 22 million.

The total equals 4 124 million people or two-thirds of the earth's population. Just to maintain, region-wise, the supply of 17.2 kg of fish per capita (1995 average), the Asian Region would require more than 70 million mt of food fish in 2020, to be supplied from capture fisheries and aquaculture. A look at the situation by region would show differences, but all the same, the indications are for aquaculture to increasingly fill the fish supply requirements of the populations.

In China, aquaculture already contributes more than 60 percent of the total fisheries production and, in 1995, the per capita supply of fish was 22.2 kg, implying a need for 32.2 million mt by 2020 assuming a constant per capita fish supply. If the same supply proportion were to be maintained, 19 million mt would be needed from aquaculture. Southeast Asia's average per capita supply is now 24.7 kg a year; with a population of 655 million it will require 16.2 million mt of food fish. In assuming that 30 percent is to be contributed by aquaculture, almost 5 million mt of food fish need to be supplied by aquaculture.

South Asia will need a total of 8.1 million mt of food fish. In 1997, only 2.33 million mt were produced by aquaculture, 97 percent of this being food fish. If the same ratio is maintained, the South Asian Region will need 7.8 million mt by 2020.The assumptions, and forecasts based on the assumptions, are summarized in the Table 3.

With a total requirement of 35.7 million mt of cultured food fish forecast for 2020, the concern for supply is evident. Since the wild catch is stagnating and forecast to slow even further, aquaculture will have to increase its contribution beyond its current levels.

The Food and Agriculture Organization of the United Nations (FAO) predicts that global aquaculture is likely to show sustained growth in the medium term and could attain 35 to 40 million mt of finfish, crustaceans and molluscs by the year 2010 (FAO, 1998). What this analysis clearly points out is that there is now a burden on aquaculture to meet the future demand for aquatic products, especially that the continued capacity to sustain further productivity increases in all the food production sectors is facing the following threats (UNDP, 1998): losses in genetic resources and biodiversity (e.g. 4 percent of fish and reptile species are threatened with extinction); loss of aquatic habitat such as coral reefs and mangroves, and severe depletion of commercial fish stocks (25 percent of fish stocks for which data are available are either depleted or in danger of depletion, and another 44 percent of fish stocks are being fished to their biological limit).

Aquatic resources are not unique in facing such degradation, as 9 million ha of land are extremely degraded, with their original biotic functions fully destroyed, and 10 percent of the earth's surface is at least moderately degraded, and the availability of safe and clean water has declined steeply to 60 percent (in 1996) of its 1970 value (UNDP, 1998). These trends, which show no signs of changing, indicate that future production of food from aquaculture will face significant challenges in access to and use of resources.

Another crucial point that should be emphasized is that the countries in this region have the highest population densities in the world. There are many issues that aquaculture has to face, but the fundamental one for Asia is that of supply and demand, which has been illustrated above. The critical concerns identified for resolution of this issue relate to social, economic, biological, environmental and institutional questions, which must be resolved with a minimum of conflict and most efficient use of resources. To put this into a positive light, aquaculture must be developed in a spirit of social harmony, environmental sustainability and economic progress.

Trends and Issues

International Trade in Aquaculture Products

Information on the trade in aquaculture products is rarely separated from trade in fisheries products, making it difficult to analyse trends in the trade of aquaculture products. Comments here are therefore based mainly on fisheries products (FAO/RAP, 2000). By volume, Asia is a net importer of fishery products, mainly because of Japan's trading patterns where an annual trade deficit (exports less imports) of more than 3 million mt has been registered since 1994. If Japan is excluded from the trade data, the developing countries of Asia would register a surplus of only 35 000 mt a year in the triennium up to

19979. The major net importers (with volumes over 100 000 mt) of fish are China, followed by the Philippines and Malaysia. The major exporters with volumes greater than 100 000 mt are India, Indonesia and Thailand. Of these, aquaculture production probably makes the biggest contribution in Thailand.

In terms of financial value, the countries with developing economies are responsible for more than half of the region's fishery exports, and there is a dramatic change in the pattern of imports and exports. Exports from developing countries in Asia are mainly high-value products that are directed primarily to developed economies. In the triennium ending in 1997, Thailand was the leading exporter in the region, exporting US$2.92 billion annually, a significant portion of which was cultured shrimp. Other major exporters with more than US$1 billion of fish products per year over the triennium include China, India and Indonesia. In terms of financial value, the countries with developing economies are responsible for more than half of the region's fishery export trade, which is directed primarily to developed economies. Between 1993 and 1996, Thailand was the leading world exporter of fish products, at annual values of around US$3.4 billion (although displaced by Norway's fishery exports in 1997). Japan is the leading importer of fish products in the world, with a value of US$15.5 billion in 1997. United States absorbs about 10 percent of the total value and these two countries and the European Community (EC) (including the value of the intra-EC trade) import 75 percent (in value) of internationally traded fishery products.

Shrimp Trade

In financial value, shrimp is the most important traded aquaculture commodity in Asia, and Thailand is the world's main supplier of cultured shrimp, with an output of over 200 000 mt. In 1997, the shortage of shrimp on the world market was acute; but Asia, as a whole, was able to maintain its share of the United States market, and smaller exporting countries in the region such as Indonesia and China performed well, although China has of late rarely exported any shrimp.

Indian shrimp exports to Japan increased by 6.6 percent to a record 59 100 mt, while total exports in the late 1990s have exceeded 100 000 mt, of which around half were cultured. The reasons for increased import to Japan included EU restrictions on Indian seafood, which started in August 1997.

In 1997, Japanese imports of shrimp fell by 7 percent to 267 200 mt, the lowest figure in nine years, reflecting a downward trend that was apparent throughout 1997. The United States shrimp market was very strong, owing to the country's expanding economy and to the high value of the US dollar, and prices rose by 20 percent in one year. United States imports expanded by 10 percent in 1997, overtaking Japan as the world's major shrimp market for the first time. Within the region, for markets such as China (including Hong Kong), Singapore and Malaysia, the imports of fish and fishery products for domestic consumption continued to be lower in 1997 than in 1996 due to the economic problems experienced.

Seaweed Trade

Seaweed exports bring significant earnings to the Philippines, China, Indonesia and the Republic of Korea. The Philippines produced over 620 000 mt in 1997 of mostly Eucheuma seaweed, from which the colloid carageenan, which is the export product, is extracted.In East Asia, seaweed (Laminaria and Undaria) are also produced and particularly Undaria, exported.

Mollusc Trade

Mollusc exports from the region, including intra-regional trade, increased from 240 000 mt in 1988 to 335000 mt in 1997, peaking at 380 000 mt in 1996. Important exporting countries are China, Democratic Peoples Republic of Korea, Republic of Korea, Malaysia and Thailand. The predominant products are clams, scallops and oysters.

Finfish Trade

The trade in live marine fish, particularly reef fishes, has been growing steadily, spurring work on their aquaculture. Market analysis done in 1995 indicated that the total seafood market in Hong Kong and southern China, the main markets at present, was over 220 000 mt a year, within which the estimated annual demand for high-quality live reef fish was 1 600 to 1 700 mt. This study forecast compound growth rates of more than 12 percent, in other words, doubling every six years.

In 1997, Hong Kong consumed 28 000 mt of live fish, of which groupers represent 35 percent by weight and 50 percent by value. Tilapia, filleted or chilled, is a growing export species. Taiwan Province of China supplied half of the 50 000 mt of tilapia consumed by the United States in 1998. Other exporters of tilapia to the United States market, but as yet in small quantities, are Indonesia and Thailand.

Singapore is also a significant market for high-value marine fish, although local cage farming has mainly satisfied its domestic consumption. To meet expected growth in demand, it has initiated the move to farm marine fish in offshore deep-water cages, targeting annual production of 40 000 mt that would be supplied with fry from local hatcheries.

The breeding and growing of ornamental fish and aquatic plants is also a growing industry. The value of world trade (wholesale and retail) in these commodities has been estimated at US$5 billion, generating employment and income to breeders and growers. Singapore, Malaysia and Thailand have well established and mature industries. Countries such as Sri Lanka and the Philippines have discovered this sector fairly recently, seeing that the activity offers opportunities for rural families to earn income from cultivating freshwater ornamental fishes.

Discussion

As the Asian economies develop and become more affluent, opportunities for intra-regional trade will increase. Before the economic downturn in the region in the 1990s, intra-regional seafood trade was increasing steadily. Some countries, such as China, Japan and Singapore, continue to be significant importers of seafood within the region and, as other economies recover, these will again play a more important role in trade within the region.

As more countries within the region develop their economies, and as costs will rise in all likelihood, changes will occur in aquaculture production costs. Where the products are destined for export, trade privileges that have been accorded to specific regional countries may also change.

The recent withdrawal, by the European Union (EU], of trade privileges for seafood exports from Thailand to the European Community (EC), for example, has had a significant impact on the export patterns for shrimp in particular and, hence, the cost structure of that industry.

Such events will lead to increasing competition between countries within the region for access to key markets. Countries with lower production costs will take a larger share of regional exports, either directly or via third countries. This phenomenon has already occurred for certain products within key international markets, and since most countries in the Asian Region are producing and exporting within similar markets, price competition for some commodities can be strong.

Cultured shrimp, in particular, has come under threat in key export markets due to adverse publicity concerning the environmental and social impact of some aquaculture activities. Some of this publicity maybe justified, serving to highlight some of the problems in the sectoral development of aquaculture. Nonetheless, this has the unfortunate effect of tarnishing all aquaculture activities in the eye of the consumer. Efforts to provide a reasoned, balanced view of aquaculture activities and the promotion of the benefits of aquaculture products, grown in a responsible manner, are necessary.

Changes in market characteristics (consumption patterns, prices of inputs, market saturation)

Domestic regional changes will have a significant influence on the species promoted for and produced by aquaculture. China, for instance, is targeting increasingly "fine'" (or high-value) species, responding to the rising demands of a more affluent population. On a more fundamental level, increasing populations are prompting higher productivity and, hence, more intensive culture systems.

A major impact seen within this context is that China has become a major importer of fishmeal for livestock and fish feeds, as well as most other countries relying increasingly on artificial feed formulation (i.e. carp farms in China and milkfish in the Philippines). Between 1988 and 1997, the import of fishmeal within the region increased by 66 percent (APR 8 percent) to reach 2.2 million mt. During the same period, the value increased by 87 percent, reflecting a rise in the value/mt of 12 percent. In remembering that not all fishmeal is used for aquaculture, major importers (>100 000 mt/year) are China, Taiwan Province of China, Thailand, Indonesia and the Philippines.

Economic growth, improved living standards and the higher purchasing power of local populations have seen an increased demand for food of improved quality and that is better packaged.

Inevitably, the "cost of compliance" with these demands will be imposed on the sector. For export, this situation favours countries with a well-developed infrastructure for handling and processing to the standards required, but will impose additional costs on countries wishing to develop or increase exports. However, the higher costs of processing in target export countries have led to an increase in the manufacture of value-added products in the producing countries.

The international marketing of Asian seafood has mainly been generic, with little attempt to develop brands or build a specific image. The use of brands or specific identifiers is much greater in developed countries, leading to a higher degree of awareness on the part of the consumer. Countries with large, integrated aquaculture producers (e.g. such as are found in Latin America or Europe) are increasingly using such marketing tools. Efforts leading to the environmental certification of seafood and/or eco-labelling will also increase this trend.

The structure of the Asian seafood export industry, particularly for those countries depending on small-scale aquaculture, incurs different levels of awareness of the present and future importance of such mechanisms in the international market. This situation could lead to Asian exporters being out-competed by those from countries that have better developed marketing "capacity".

Quality of Aquaculture Products

The importance of aquaculture production in Asia and in particular that of freshwater fish, most of which are sold in the domestic market, has a significant bearing on the concerns of human health associated with products from aquaculture. The most common method of producing low-value freshwater fish in developing countries, especially in low-income food deficit countries (LIFDCs), is to use a combination of fertilizer application and supplementary feed. Polyculture of species that have complementary feeding habits is a general practice, especially with major and Chinese carps. However, the patterns of production are changing rapidly, with a large increase in more semi-intensive farming.

There will be further emphasis on food safety issues in the future. In 1997, a WHO (World Health Organization)/FAO/NACA Study Group (WHO/FAO/NACA, 1999) looked into the food safety issues associated with aquaculture and concluded that there were considerable needs for information, recognizing that such knowledge gaps could hinder the development process. Education is needed to increase the awareness of potential hazards in aquaculture and their management. The Study Group also recognized the difficulties in applying Hazard Analysis Critical Control Point (HACCP) procedures to small-scale farming systems and that food safety hazards associated with aquaculture vary by region, habitat and environmental conditions, as well as by methods of production and management.

In Asian shrimp production and processing, food safety is given increasing attention, largely because of the quality standards required by importing countries but also because of the higher profile given to environmentally friendly practices in the aquaculture sector.

Agreements Regulating International Fish Trade

International rules and regulations play a major role in governing the trade of fishery, affecting aquaculture products in both developed and developing economies. Several major international agreements and codes of conduct include sections that relate to aquatic animals and their products. In the Asia-Pacific Region, however, there is a wide difference in knowledge and expertise between different countries Furthermore, some countries in the region are members of organizations such as the Office International des Épizooties (OIE) and the World Trade Organization (WTO), and comply with the terms of such agreements, while others who are not signatories are obviously not bound by the terms of such agreements.

Regulatory and Management Frameworks for Aquaculture

Agencies Responsible for Aquaculture Development

In Asia, the primary responsibility for aquaculture production usually resides within one government ministry or department. By contrast, the management of the resources upon which aquaculture depends, particularly water and land, is normally the responsibility of other ministerial departments. This situation inevitably causes the potential for conflicts between different departments, and confusion or lack of clarity in the application of policy.

The increasing importance of aquaculture strongly argues for governments to give priority to developing clear, well-formulated, implementable and realistic national and local policies for aquaculture development, based on financial, social and environmental sustainability. As the private sector is the key to successful and sustainable aquaculture development, the views of industry should be taken into account in respect of policy formulation, research and development.

Participation

The adoption of a "participatory approach" is receiving increasing attention in both aquaculture planning and management. Experience with coastal fisheries management has shown, in general, that the failure to include coastal residents in natural resource management can lead to a lack of community compliance, resulting in resource depletion and conflicts.

This situation can be particularly acute when the government's capacity to enforce laws and regulations is limited.

A promising approach to this problem is "co-management", which involves the cooperation of the local community in both establishing and enforcing local management rules, with the requisite support from government. The co-management approach has proved useful for community-based management of some coastal capture-fishery resources, but the devolution of ownership and management of resources towards local people and communities should be explored for coastal aquaculture. The need to develop effective local arrangements for management of aquatic resources will also be stimulated further in some countries by the trend towards devolution of decision-making powers to local government bodies, as in the Philippines, Thailand and Viet Nam.

It is also apparent that the views of non-aquaculture groups need to be considered when formulating policy and implementing aquaculture development; without such views and contributions, some problems relating to conflicts of interest may be too difficult to fully resolve.

Such issues are quite new, but there is evidence of increasing involvement of various aquaculture interest groups, including the private sector, local farmers and others, in aquaculture policy formulation. In some countries, the government and farmers have entered into discussion with NGOs and non-aquaculture groups for the formulation of management strategies. Examples have been seen in Sri Lanka, where NGOs have participated in aquaculture policy discussions, and in Malaysia, where input from various government, farming, academic and NGO interest groups was provided in the formulation of the shrimp and marine fish "Codes of Conduct".

Legislation

In many countries, aquaculture legislation is poorly developed and frequently consists of a few articles contained within a law regulating capture fisheries. During the last few years, interest has grown to develop a comprehensive regulatory framework for aquaculture whose goals would be to protect the industry, the environment, other resource users and the consumer, and to clarify rules for resource use.

This consideration is driven by a variety of factors that include the following:

* greater political attention to the sector as its economic importance and potential become more apparent;
* realisation that inappropriate laws and institutional arrangements can severely constrain the sector's development;
* evidence of environmental damage and social disruption as a result of the rapid and largely unregulated expansion of shrimp farming; and

* a growing emphasis on examining production methods as a means of improving the quality and safety of aquaculture products.

While capture fisheries are generally regulated by a single government department or ministry, aquaculture is frequently regulated by many agencies under a variety of laws. In this circumstance, the formulation of a comprehensive regulatory framework for aquaculture is legally and institutionally complex. Typically, the tasks involve drafting or amending legislation that addresses a variety of issues such as land use planning and tenure, water quality and environmental issues, fish movement and disease, pharmaceutical and chemical use, food quality and public health.

It also requires clarification of responsible authorities and the establishment of the arrangements to ensure cooperation and coordination between many different institutions that have jurisdiction over the item or issue concerned.

International Developments

The adoption of the Code of Conduct for Responsible Fisheries (CCRF) by the 1995 FAO Conference (FAO, 1995) promises to provide a significant influence on the development of regulatory systems for aquaculture in the coming years. Article 9 of the Code deals with aquaculture development and sets out a wide range of relevant principles and criteria. In addition, the "Jakarta Mandate", which was adopted by the second Conference of the Parties to the Convention on Biological Diversity in 1995, provides useful guidance regarding aquatic biodiversity and related environmental aspects that should be taken into account when developing coastal aquaculture.

During 1994 and 1995, several regional workshops10 were organized to examine environmental, legal and institutional issues associated with aquaculture. Recently, FAO published a compendium on aquaculture and inland fisheries legislation in the Asian Region (FAO, 1996). Although new national laws to regulate aquaculture comprehensively may be desirable in many countries in the region, other options are now being explored. This is because the time required to develop and pass new comprehensive legislation is likely to be several years, while the rapid development of the sector has created an urgent need for regulation. These options include the enactment of regulations that exist under existing legislation and the application of voluntary approaches such as guidelines and codes of practice.

Creating incentives to encourage compliance or disincentives for noncompliance or failure to join the scheme can enhance the effectiveness of voluntary codes of practice. One example is the policy that was adopted in India in 1995 by the National Bank for Agriculture and Rural Development (NABARD), which was a major source of refinancing for brackishwater prawn farming in the states of Andhra Pradesh and Tamil Nadu. This policy sets out the conditions to be met before NABARD will provide credit to banks for aquaculture developments. Similarly, Malaysia is considering the requirement that aquaculture enterprises which are given preferential "pioneer" tax status must adhere to a code in order to retain the various tax incentives accorded. Clearly, there is scope for increasing the use of such conditional public and private-sector funding to encourage the application of sustainable aquaculture practices.

The Australian State of Tasmania has recently passed legislation that provides another interesting example of what can be done to address this issue. The new laws (notably the Marine Farming Planning Act [1995] and the Living Marine Resources Act [1995]) impose that the preparation of marine farming development plans must cover areas, rather than sites, and provide for broad community participation in the preparation of these plans. An environmental impact assessment must also be done, with the establishment of a marine farming zone, before site leases are granted to marine farms.

Malaysia is developing an integrated regulatory system for aquaculture, in the short to medium term, but without formulating a specific aquaculture act. Instead, new regulations will be incorporated under the existing Fisheries Act, and a voluntary Code of Responsible Aquaculture Practices will be introduced for cage culture and shrimp farming. These measures will be supported by incentives, and institutional structures will be strengthened to ensure the effective monitoring and continuing formulation of aquaculture policy.

Many regulatory systems could be improved by including mechanisms that are designed not only to prevent or reduce the risk of environmental harm but also to help right any damage that may occur. The Tamil Nadu Aquaculture (Regulation) Act of 1995 not only set out conditions to improve the siting and management of aquaculture facilities, but also established an "Eco-Restoration Fund", funded partially by deposits from aquaculturists, to be used for remedying any environmental damage caused by aquaculture farms. In addition, the constitution of the Aquaculture Authority of India has been made for the regulation of shrimp aquaculture.

Environmental Aspects of Aquaculture

Environmental issues have become of increasing concern for several reasons. The first is recognition of the increasing pressure on resources in some coastal and inland areas. More attention has been given to the impact of aquaculture on the environment in recent years, partly induced by some well publicized "crashes" within the shrimp industry.

This has been accompanied by the publicity given to environmental and social issues surrounding aquaculture. This type of exposure has had a profound impact on the public's perception of aquaculture, changing it from the "blue revolution" which would improve the availability of cheap, affordable protein to poorer people, to that of being an environmentally unsound means to produce luxury food items for consumers in developed nations.

Australia is adopting integrated farming technologies in a watershed where water is an economic resource and is at a premium (i.e. use must be paid for). This situation is in contrast to that encountered in some countries in the region where traditional integrated farming practices are reducing their levels of integration, applying less species diversification and more emphasis on high-value products.

Balanced Resource Use

There has to be a balanced and more efficient use of resources. Obtaining conflicts. This situation can be particularly acute when the government's capacity to enforce laws and regulations is limited.

It appears likely that economic issues related to resource use and the "polluter pays" principle will be faced, in the next few years, by aquaculturists in some countries of the region.

Minimizing Negative Impacts

Guidance is emerging for the application of practices for improved environmental management, and strategies have been identified for the following:

* Technology and farming systems: In recognizing the importance of appropriate farming technology/ system and management of inputs and outputs, special attention is given to the major resources used (i.e. feed, water, sediments and seed). At this level, management actions mainly involve farmers and input suppliers.
* Adoption of integrated coastal area management approaches: The importance of integrating aquaculture projects within existing ecological systems in coastal areas is increasingly recognized. This approach requires the consideration of proper site selection and the application of planning and management strategies that allow the allocation of resources among different users.
* Policy and institutional support: It is necessary to have a clear and supportive policy framework for aquaculture. Particular issues include aquaculture legislation, economic incentives and disincentives, actions for the public image of the activity, private-sector and community participation in policy formulation, and increasing the effectiveness of research, extension and information exchange. Policy decisions and their implementation play a strong role in influencing the management possibilities at both the farm and local area levels.

The environmental assessment of an aquaculture area, rather than targeting individual projects, is now being used to identify opportunities for minimizing the impacts of the aquaculture sector within a particular area (e.g. in a bay, an estuary or a watershed). Such applicable measures are promoted through a wide range of instruments and should be brought together within the framework of an aquaculture development plan, ideally as part of an Integrated Coastal Management Plan.

There is a growing awareness that environmentally friendly aquaculture makes good business sense, and this is particularly true for commercial producers. It is increasingly important when considering the perceptions of target export markets and provides an incentive for both the industry and supporting governments to further advance the adoption of environmentally and socially responsible farming practices through appropriate standards or codes of conduct. Recognized codes of practice require further development and implementation and are becoming particularly important for aquaculture products that are traded internationally.

Aquaculture and Environmental Rehabilitation

It is well known that inland aquaculture, particularly that within integrated agricultural farming systems, can provide many positive environmental benefits (e.g. water storage on small-scale farms, efficient use of farm resources). Perhaps less well known and publicized is the fact that coastal aquaculture, including shrimp culture, can also contribute to environmental improvement. For example, seaweed and mollusc culture can contribute to the removal of nutrients and organic materials from coastal waters. It can also provide an alternative source of income and employment for people involved in environmentally destructive fishing practices and other forms of habitat destruction. Other activities of note include the following:

* Mixed aquaculture-mangrove systems are being used to restore mangrove habitats in some countries (e.g. Indonesia's tambaks or earthen ponds, Viet Nam's mixed farming systems);
* Coral reef fish mariculture provides an effective alternative to destructive fishing practices in coral reef areas;

* The rehabilitation of fish populations through stock enhancement; and
* Aquaculture itself is also a technique for the effective monitoring of environmental conditions.

A recent review has suggested that aquaculture within saline inland areas can provide benefits through the use of previously degraded resources, contributing to environmental rehabilitation and poverty alleviation among communities whose livelihoods such damaged resources.

Aquaculture is also being integrated as a component of mangrove rehabilitation projects (as in the Mekong Delta) and within coral reef management activities (small-scale cage culture).

Biodiversity, Genetics and Aquaculture

The potential impact of farmed aquaculture stocks, especially introduced exotic species, on wild fish and fisheries has been highlighted at many international meetings. Among the potential impacts identified are a loss of genetic resources within healthy or undisturbed populations and a dilution of the wild gene pool through interbreeding with escapees from aquaculture facilities. Initiatives have been made recently to breed indigenous species for stock enhancement purposes as well as for aquaculture, but these are new initiatives and results that could guide development work are not available yet.

Aquafeeds and Feeding Strategies

When feeds are used in aquaculture, their contribution to production costs varies following the species and the system employed but can be as high as 50 percent of the total costs. Following concern as to the sustainable supply of fishmeal and fish oils, research has focused on identifying alternative, economically viable sources of proteins and lipids for incorporation into feeds. Improved management for more efficient feed use is another area of interest.

Some research (i.e. in Hong Kong) has taken a step further towards the development of environmentally friendly feed. Little information is available on the production of aquaculture feeds in Asia, but a recent estimate indicates a supply of some 1.9 million mt in Asia (excluding China), where 1.2 million mt were for fish and 0.7 million mt for shrimp. An estimate for China (1996) indicated a total aquaculture feed production of 5 million mt, but it was unclear if this figure also included farm-made feeds. The future prospect for fishmeal and oil for animal feeds is that global supply is expected to be stable (FAO, 1998).

A distinction must be drawn between the different feeds manufactured for aquaculture species, i.e. whether for high-or low-value species. The higher values of commodity species, or those reared for export, can justify the investment required for the manufacture, distribution, purchase and use of specially formulated feeds. However, much of Asian aquaculture involves smallholder-based production of low-value species, either for personal consumption or for local sale. In the latter case, the input cost of feeds is generally negligible and, if feeds are used, they usually consist of cheap, readily available materials.

International Development Aid to Aquaculture

While Asia has received much of the international external assistance for aquaculture development, FAO (1998) has reported reduced levels in recent years. However, the recent trend for such support has been through assistance given to projects having a wider scope. These include as socially oriented projects, environmental rehabilitation projects, as well as integrated area development programmes, such as the Mekong River basin development programme, and the Bay of Bengal Programme (BOBP). In many of these projects, aquaculture is a component within a broad-scale development.

For the period 1988-1995, the Asian Region accounted for 65 percent of commitments made by funding agencies and donors to development assistance (which is different from actual disbursements) and 38 percent of the projects, while Africa was the recipient of 16 percent of commitments and about a quarter of the projects. The major beneficiaries during this period (i.e. countries receiving at least 1 percent of total aid) included India, China, Bangladesh and Mexico, which all received major loans from development banks, accounting for about 64 percent (about US$638 million) of external aid to aquaculture.

The two most common sources of credit for Asia's developing countries have been the World Bank (WB) and the Asian Development Bank (ADB). In 1997, WB provided US$409 million to South Asia's agriculture sector (a decrease of 26 percent compared to 1995), while the Asian Development Bank approved 632 loans, of which 10 percent went to the agriculture sector and 50 percent to the financial sector.

Other multilateral programmes that assist small farmers are the United Nations Development Programme (UNDP) (for agricultural development, including training and provision of experts) and the United Nations Development Fund for Women (UNIFEM) (provision of financial capital enabling women to establish small businesses, including aquaculture).

The United Nations Children's Fund (UNICEF), the United Nations Drug Control Programme (UNDCP) and others, including FAO, extend support to small farmer development programmes under various schemes (Technical Cooperation among Developing Countries (TCDC), Special Programme on Food Security (SPFS), Technical Cooperation Programme (TCP)). In addition, a number of donor agencies provide small-scale enterprises with development fund assistance. These include the United States Agency for International Development (USAID], the Australian Agency for International Development (AusAID), the Canadian International Development Agency (CIDA), the Danish International Development Agency (DANIDA), the Swedish International Development Agency (SIDA), the Norwegian Agency for Development Cooperation (NORAD), the Deutsche Gesellschaft für Technische Zusammenarbeit (GTZ), the Department for International Development (DFID, United Kingdom), the Groupe de Reflexion d'échange Technologique (GRET, France) and agencies from Belgium and the Netherlands. Assistance to research and development is also provided by the Australian Centre for International Agricultural Research (ACIAR), the Danish Cooperation for Environment and Development (DANCED), the Japanese International Cooperation Agency (JICA) and the other afore-mentioned agencies.

Assistance by Banks, Private Sector to Aquaculture Development

Private-sector investment for aquaculture development has tended to be limited to the culture of high-value species, where the systems are more capital intensive. The degree of personal involvement of the investor depends upon the scale and cost of the project. Small-scale projects may be financed directly by the owner or jointly with other investors, with the farm operating without recourse to external sources of finance. As the investment and operating capital requirements increase, it is often necessary for such entities to turn to external sources of funding (e.g. local or national banks).

For larger scale production units, particularly in cases where the investor has little or no direct control over day-to-day operations, such investment is often assessed alongside other potential opportunities. These circumstances can be quite volatile and depend on the relative performance of the investment(s). For example, the short-term nature of many of the shrimp farming investments has resulted in a lot of movement of investment capital, both into and out of the sector. The Asian Region does not possess a well-developed venture capital market that is willing to invest in new projects of potentially high financial risk. This situation severely limits the region's capability to develop and bring to the market new technologies without recourse to foreign investment.

This situation can also lead to a loss of opportunity, particularly where the fruits of research that have been achieved in the region benefit projects and investors elsewhere. Additionally, future prospects for private-sector financing are likely to be affected by the current weak economic situation in much of the region.

It is clear that many projects that were funded in the years before the economic crisis in Asia were not subject to the kind of scrutiny that would have identified high-risk investments. The banking reforms that are being undertaken in the region are likely reduce the level of financing available to aquaculture projects unless these can be demonstrated to be technically and economically viable. At the same time, the increasing costs of compliance with government regulations, as well as the trade agreements referred to, will increase the cost of entry into the commercial aquaculture sector.

NGO support to projects involving aquaculture has also increased, noting that there are thousands of NGOs in Asia. National development banks have tended to give a higher priority to projects producing high-value crops, although some, like India's NABARD, the Grameen Bank in Bangladesh, Bank Pertanian Malaysia, Bank Rakyat Indonesia, Bank of Agriculture and Agricultural Co-operatives of Thailand, and the Land Bank of the Philippines and others have developed loan programmes for small projects oriented towards rural development.

The private sector has played a major role in Asian aquaculture development, although principally for commercial rearing of higher value species. Due to the predominance of small-scale farms in some countries and the rate of development of intensive aquaculture, feed companies and other suppliers have proved to be a significant source of information for farmers. Shrimp feed companies, in particular, frequently provide information and technical support as part of their customer services. Such information is usually aimed at increasing sales.

Sometimes, this can lead to the widespread recommendation of inappropriate practices, such as overstocking. On the other hand, it has been customary to organize seminars, with invited independent experts, to pass on new information to farmers. On balance, the impact of this type of information transfer has been positive for the production sector. Most feed companies also provide technical support services for both their current and prospective customers, which may include checking water quality, fish or shrimp health and providing advice on the farm. Some suppliers have introduced testing services for shrimp viruses.

The creation of farmers' groups has become more common in recent years, partly as a response to the pressures put on the production sector by external sources, but also with the realisation by the farmers that a strong professional lobby was needed to put aquaculture on the political agenda. This type of action has improved the exchange of views between the appropriate government bodies and the private sector and has also improved the awareness of wider issues among the members of such farmers' groups.

Within the region, much aquaculture investment has been, essentially, of a short-term nature. This position, when combined with the lack of local sources of venture capital, has meant that the region, with one or two exceptions, lags behind considerably in the private-sector development of new technologies. An exception to this was the establishment of a consortium of several private-sector companies for the development of a shrimp-breeding programme in Thailand. However, this might not have been established without the major involvement of a government agency that is a major shareholder in the venture. This position demonstrates the contrast with other shrimp breeding initiatives in the Americas that are mostly financed with private-sector funds, albeit using research results achieved through governmental support (universities, institutes and other public bodies).

5

Regional Institutions and their Assistance

Apart from NACA, there are a large number of organizations and agencies that are active in aquaculture development in the region.

They include the Asian Institute of Technology (AIT), The Southeast Asia Fisheries Development Center (SEAFDEC) (through its Aquaculture Department), the International Center for Living Aquatic Resources Management (ICLARM) (which has a sizeable research programme in the Asia-Pacific Region), the Aquatic Animal Health Research Institute (AAHRI) (a regional resource centre for fish health management and now a reference centre on epizootic ulcerative syndrome (EUS), INFOFISH, the Mekong River Commission (MRC), and the Asia Pacific Economic Co-operation (APEC) (which is collaborating in a study on manpower development for aquaculture and the grouper R & D Network). Information is shared among the organizations and most are managing a number of specific databases for development and management purposes.

There is a growing trend towards collaboration that focuses efforts and links or integrates different initiatives, providing wider attention to and a greater impact on development issues and problems. This approach makes better use of the available resources and is also opening opportunities for governments to decide unified positions, hence a stronger voice in relation to certain international issues, trends or agreements.

One example is a project concerning the responsible movement of live aquatic animals11 that has enabled access, collectively by 21 Asian governments, to the capabilities and assistance of OIE and FAO for promoting quarantine, health certification and a responsible approach to fish health management. Successful work would have a positive impact on both farming practices and trade. Likewise, cooperation between WHO, FAO and NACA on food safety issues has brought regional attention to the concepts and practices of managing food safety hazards that are associated with aquaculture products, an extremely important consideration for consumer protection, trade and stimulation of better management practices.

For manpower development, many national and regional institutions and agencies provide tertiary education. Vocational and short-term technical training provide valuable support to human resources development in the region. While NACA has a large short-term training programme, a number of regional institutions, such as AIT, provide both long-term and short-term training programmes. The regional centres of NACA in India and China have expanded their cooperation with tertiary education institutes to be able to provide longer term degree-based programmes.

If aquaculture is to sustain its growth momentum within the region more effort must be focused towards the building of human capacity. To be able to respond to and to meet the increasingly complex development challenges faced by aquaculture, this effort must be directed at all levels, from the farm worker to the policy maker. At the present, a proposal for a potential cooperative regional strategy to meet these needs is being prepared within the cooperative aquaculture education project of APEC/NACA/Deakin University.

Major Shifts, Changes in Regional Policies that Influence Aquaculture Development

The two major influences are the broadening of the aquaculture development objectives from economic to social, which focuses on poverty alleviation, livelihood development and food security, and the links made between sustainable aquaculture practices and trade.

Poverty Alleviation and Food Security

Throughout the region, the expressed national goals and priorities concerning aquaculture have invariably included and even emphasized social goals. The key issue concerning this topic has been in respect of the emphasis given to commercial and/or export-oriented aquaculture, on the one hand, and to the small-scale food security oriented activities on the other. Both the levels and the balance of the support given to either subsector are part of this issue. It has to be said that increasing recognition is being given to the importance of small-scale, socially oriented aquaculture. Initiatives have been made recently at the regional level to focus the attention of both government and regional organizations on this issue.

ENVIRONMENT AND TRADE

The role of the fish farmer is changing from merely that of a producer to being an integral part of a chain for the production and delivery of high-quality food products to the consumer. Hazards can be introduced within the chain at the production stages on the farm, and these can be spread during processing and preparation. The means of assuring food safety can be difficult to determine when certain microbial hazards can cause diseases in humans but not in fish, as is the case for some naturally occurring pathogens.

While the implementation of safety assurance programmes is well advanced in the fish-processing sector, the application of such programmes at the fish-farm level is a new approach. The application and enforcement of HACCP regulations at the production level could cause significant difficulties for Asian farmers unless major efforts are made to develop technologies and methodologies for the reduction of potential risks.

PRODUCTION PRACTICES AND TRADE

The link made between production practices and their impact on both the environment and trade has been reflected in recent regional and national programmes. Some of the measures that have been started include environmental impact assessments, the development of guidelines to implement the FAO Code of Conduct for Responsible Fisheries (CCRF) and the development of policies on zoning and integrated coastal zone management (ICZM). These measures are being developed and achieved seriously and in a concerted manner.

Impacts of the External Environmental on Aquaculture

These may be positive or negative. Nutrient enrichment of water bodies may provide nutrients that are beneficial to aquaculture production in some extensive culture systems (e.g. seaweeds, molluscs). However, excessive loadings with urban and industrial wastes can have severe consequences for aquaculture operations, particularly shellfish culture, when exposed to contamination by toxic pollutants, pathogens and phycotoxins. Contamination of coastal waters with phycotoxins associated with plankton blooms has caused major losses in East Asia in recent years (Hong Kong and China, in particular).

With increasing aquatic pollution and the physical degradation of aquatic habitats by other non-aquaculture developments, aquaculturists face the risks of mass mortalities of farmed stock, disease outbreaks, product contamination and reduced availability of wild seed or broodstock.

Impacts of Aquaculture on the Environment

Many types of aquaculture can contribute positively to environmental improvement. The recycling of nutrients and organic matter through integrated farming systems has long been recognized as being environmentally sound and is still widely practised in the region. Recent developments in Integrated Pest Management (for example, through NGO interventions in Bangladesh) have shown how rice-fish culture can help farmers reduce the use of environmentally damaging pesticides. The use of wastewaters for freshwater aquaculture, mollusc and seaweed farming can recover excess nutrients, thereby reducing risks of eutrophication. In programmes for the restoration and recovery of endangered fish species and stocks, hatcheries and aquaculture systems have been used both to provide a temporary sanctuary and to increase numbers of individuals for re-introduction into the wild.

Negative impacts have been associated mainly with intensive and monoculture systems, the effects of which can include nutrient and organic enrichment of recipient waters, resulting in a build-up of anoxic sediments, changes in benthic communities and the eutrophication of lakes. In some areas, large-scale shrimp culture has resulted in degradation of wetlands, localised water pollution and salination problems. The misapplication of chemicals, the collection of seed from the wild, the introduction of exotic species and the overuse of fishery resources as feed inputs are topics that have also raised concern in some locations.

Such problems have focused attention on the development of appropriate management strategies that, in some cases, have had trade implications for aquaculture products (mainly shrimp) traded on international markets. Thailand, for example, is now looking into a "certification" process for marketing shrimp aquaculture products that are produced in environmentally sound ways.

Impacts of Aquaculture on Aquaculture

This intra-sectoral issue has raised quite specific problems that have been particularly acute in intensive cage culture and shrimp culture. These concern self-pollution and the transmission of diseases that have occurred in areas where the high density of farms requires the use of water that has been contaminated by neighbouring installations. In some cases, significant losses of farmed stocks have occurred, significantly affecting financial returns. Encountering self-pollution problems at least provides farmers with very strong incentives to improve environmental management. Indeed, several regional countries have made significant changes to their management practices designed to reduce disease risks, which have apparently led to reductions in negative environmental impacts. Examples include the closed and recirculation shrimp farming systems.

Other Interactions with Aquaculture

There are a variety of interactions between aquaculture and other sectors, which include capture fisheries, agriculture, forestry (including coastal mangrove forests), industry, tourism and navigation. As in the case of environmental interactions, these may also be positive or negative.

In Malaysia, there is a growing interest to integrate some aquaculture operations with tourism (e.g. marine cage culture and put-and-take fishing ponds in Sarawak). Much inland aquaculture is carried out in agricultural areas and is, therefore, well integrated into agricultural farming systems, adding value to the total farm output and, in some cases, making best use of on-farm resources.

In some coastal areas, some significant conflicts have arisen. For example, social conflicts have occurred between rice farmers and shrimp aquaculturists in areas of Bangladesh. Such situations, while generally not well documented, suggest that more attention will have to be given to techniques of conflict management. Given that interactions between coastal aquaculture and other sectors are likely to increase due to rising population pressures in Asia's coastal regions, more attention needs to be given to these important issues.

Australia is moving to resolve potential issues related to aquaculture "interactions" through the integration of sectoral interests. The plan is to develop a policy for integrated aquaculture development, within which the objectives would be:

* to facilitate the establishment of a primary industry sector, at a national level, which would be the Australian Integrated Agri-Aquaculture Systems; and
* to provide a strategic R&D focus for integrated agri-aquaculture systems.

These considerations would apply to all relevant activities, agencies and bodies, primary producers and individuals/companies active in agriculture, fisheries and aquaculture.

Technological Advances over the Last Decade

The standard that is usually set for technological development in aquaculture is livestock husbandry and, by this measure, aquaculture is way behind. The three commodities that have received intense regional R & D attention, particularly for breeding and culture techniques, are shrimp, tilapia and carps, while milkfish, seabass and catfishes have received less attention.

Genetic improvement of aquaculture species in Asia has been most advanced for tilapias and carps. Surprisingly, despite their high value, work on shrimp genetics in Asia has been confined to academic research although, more recently, several private ventures have expressed interest in such developments. Tilapia, although an introduced species, has become widely distributed in Asia. Recent advances in selective breeding and sex control in tilapia have had a major impact on farm procedures (G. Mair pers comm.).

The topic of feed development has seen some progress, and techniques to make feeds on the farm have been widely documented, improved and disseminated.

Research on mass seed production and hatchery systems has been successful for carps, tilapia, milkfish, freshwater prawn and indigenous freshwater species. Much remains to be done for marine shrimp (particularly on broodstock development), grouper (especially mass seed production) and mangrove crab (also mass seed production).

The development of new species (particularly marine species) and aquaculture technologies is a key strategy for several Asian countries. The development and implementation of best management practices (BMP) will also require more attention, and it is considered that management of the complete farming system will become more important. A further issue is that, as a wider number of species are considered for aquaculture, research needs will increase and require more investment in R and D facilities and projects.

Techniques for disease diagnosis, as well as products and methodologies for prevention and control, have only recently been developed and applied in the field, largely because of the occurrence of viral shrimp diseases. The problems encountered have highlighted the importance of having healthy seed in sufficient quantities to support the development of aquaculture.

Quality concerns are also important in inland freshwater aquaculture, such as in carp culture, where genetic deterioration and poor quality seed have affected the performance of freshwater aquaculture, and in some other species, such as milkfish. Work on seed quality improvement has been relatively limited but, with few exceptions, clear guidelines on the characteristics of high-quality seed are available. In most cases, demand usually outstrips supply, at least on a seasonal basis. In such circumstances, there is little to no possibility of introducing more strict quality demands.

TRENDS IN PUBLIC SUPPORT TO THE SECTOR

Following the Kyoto Conference of 1976, a major change that occurred was the increase in government support to aquaculture. Regional cooperation was enhanced and organizational structures were put in place to transform aquaculture from a traditional practice to a more science-based activity. Governments in the region have recognized the role of aquaculture as an important food provider and income earner. Government planning and development increased, elevating aquaculture to a status close to that of capture fisheries.

Active steps are now being taken to encourage the involvement of the private sector in providing services to the industry. These services include the production and distribution of seed, postharvest practices and marketing, and research and development. Governments do recognize, however, that strategic services need more support, such as central hatcheries for broodstock maintenance and development and research and development programmes directed towards the small-scale aquaculture sector.

RESOURCE UTILIZATION: PATTERNS AND TRENDS

The growing scarcity of land-based resources is prompting the sector to look for alternative sites for aquaculture. Maritime countries are investigating the use of offshore waters, while other areas of investigation include derelict water bodies and wastelands, impounded water bodies and, although not new, irrigated rice lands for rice-fish culture. Increasing interest is being paid to ofshore cage culture, and a number of countries (including Australia, Singapore, Malaysia, the Philippines and Thailand) are making serious attempts to apply this system.

India, Bangladesh and Sri Lanka are improving management systems for inland open water bodies, while China has good technology for culture-based lake and reservoir fisheries. Integration in small-scale aquaculture is being promoted for greater farm productivity and to lower risks.

For a number of reasons, it appears inevitable that there will be an increase in the acreage of impounded waters in the region. Once in place, such waters need to be used for aquaculture purposes to an increasing extent, whether for culture-based fisheries, cage culture or a combination of both. Most countries recognize this to be a relatively uncontroversial way of increasing fish production through aquaculture. Culture-based fisheries have the additional advantage that the rearing process is of least environmental disturbance. Key examples can be found in China, which has substantially increased its yields from culture-based fisheries.

Main Constraints and Opportunities

Policy Constraints

* Short-term leases of the public water bodies for aquaculture development inhibits planning and development.
* Research emphasis is on short-term economic returns rather than a balance with long-term social benefits.
* Lack of or weak implementation of zoning policies.
* Lack of clear guidelines to address issues such as genetically modified organisms (GMOs), non-trade barriers, food safety, and externalities from production and resource use.

Resource Constraints

* Multiple use and multi-ownership of ponds and other water bodies.
* Limited access to public water bodies for pen and cage culture.
* Increasing pollution, shortage of feeds, disease incidence, and intensification problems (health and feed).

Institutional Constraints

* Weak institutional capabilities of aquaculture extension agencies, including government

organizations and NGOs, also weaken the transfer of technology.

* Too many government organizations and NGOs are involved in aquaculture support services, duplicating and overlapping activities.
* Limited access of the sector to commercial credit.
* Technology transfer mechanisms are weak.
* Research-related constraints, including resources, skilled researchers, facilities, coordination among institutions and workers, and poor linkage between basic and applied research and between research and production sectors.
* Weak links between research and development activities in the freshwater aquaculture sector.

Socio-economic Constraints

* High investment costs.
* Land use conflicts.

Lack of Security of Investments

* Legal constraints.
* Lack of access to resources and security of tenure or user rights.
* Lack of incentives for entrepreneurs.
* Lack of adequate guidelines for responsible aquaculture.
* Lack of unified regulations on use of resources by aquaculture.

Technical Constraints

* Absence of statistics and information on different types of water bodies and weak statistics and databases for aquaculture production by different farming systems.

* Lack of infrastructure for fish feed manufacture and distribution in different parts of the region.
* Lack of technology for the diversification of aquaculture systems.
* Inadequate manpower for research and extension services.
* Lack of well-trained manpower to manage or operate aquaculture projects.
* Inadequate and unreliable supply of quality fish seed.
* Lack of markets for some aquaculture commodities.
* Lack of technology for postharvest practices and value addition.
* Fish health management capacities are inadequate.
* Increasing scarcity and decreasing quality of water resources.

Financial Constraints

* Lack of government resources to fund development projects.
* Lack of investment guidelines for the private sector.
* General perception of aquaculture as a high-risk or non-bankable investment.

Outlook and Prospects

The issues that have been covered within this paper are the demands made of aquaculture, its ability to fulfil these demands, the threats to that ability and the opportunities to overcome these threats. These opportunities have been articulated by governments into a development plan for Asia, made at the regional workshop to formulate the Asian aquaculture development strategy for 2000-2020 conducted by NACA in September 1999. The governments envisage aquaculture to be "a major provider of food and applied to reducing societal disparities and inequities." Based on this vision, the mission will be:

* to fulfil the aspirations of all sectors of the populace, through a people-oriented approach and focusing on aquaculture for development;
* to develop aquaculture in a responsible manner and in harmony with the environment;
* to transform the emphasis of aquaculture from being an activity that is resource-dependent towards being knowledge-based; and
* to continue the pursuit of shared governmental objectives through regional cooperation and to strengthen the existing spirit and substance of this.

Development Goal

The development objective of the plan is to improve the quality of life of the people in the aquaculture sector and to help strengthen the social and economic status of the Asian countries/territories through the assurance of food security, the alleviation of poverty, the improvement of livelihoods, and increasing the net worth of the national economies.

Strategic Objectives

To achieve the mission and the development goal, the plan will pursue the following strategic objectives:

IMPROVE AQUACULTURE TECHNOLOGY

The continued development of farming systems and technology is essential and imperative. There is a need to improve farming technologies, to encourage integrated farming and crop diversification, and to increase the focus on improving efficiency and the optimal use of resources, achieving these through technological advancement and possible mechanization. Major challenges to meet this objective relate to securing and efficiently providing the inputs required for the different aquaculture systems.

Assure the Availability of Good Quality Seed

A major constraint to increasing the productivity of farming systems is widely recognized to be the inadequate supply of good quality seed. Emphasis needs to be placed on the importance of the role of local hatchery production, alongside quality certification and accreditation of seed and the development of strategies to further improve seed quality. The development of breeding and hatchery technology, genetic improvement and domestication are additional key objectives for securing the seed supply for major aquaculture species.

Improve Efficiency

The efficient use of feed and fertilizer inputs, particularly through the use of readily available local raw materials, and the minimization of environmental impacts, is highlighted; fishmeal replacement, improvements in feeds and feeding technology and the development of environmentally friendly, low pollution feeds need to be tackled. Concerns over the incorporation of GMOs as feed ingredients will also be addressed.

Improve Availability

This objective will address the development and supply of suitable incentives for aquaculture (for example, tax credit schemes) and include the creation of a more conducive environment for adequate insurance, the provision of long-term land tenure (to increase certainty for investors) and investigate ways of encouraging more private-sector participation in the financing of aquaculture development.

In terms of access to financial resources, schemes will be developed for the simplification of credit arrangements for farmers and improving the coordination between financial and technical agencies. The plan will pay particular attention to the small-scale farming sector.

Social and Economic Impacts

Diseases cause significant economic losses, and the prevention and control of disease will remain an important issue. The objective is the continued development and implementation of government quarantine and health certification measures; these will include monitoring and surveillance programmes, contingency planning and emergency responses, establishment of diagnostic capability and methodology standardization. The plan will encourage a shift from governmental health management and control towards preventive husbandry and on-farm health management. Focus will be given specifically to health measures that can be used by rural farmers.

Land and Water Resources

Such a strategy needs to address the likely reduction, in most countries in the region, of the availability of water and land resources for aquaculture. The objective is to find ways to make more efficient use of water resources and to maintain the region's lead role in global aquaculture. Appropriate systems that offer good scope for increasing the productivity of agricultural land, such as integrated farming, will be developed and promoted.

In some countries, there may be scope for enlarging the sites available to aquaculture by employing unused or under-utilized areas (e.g. inland saline areas, coastal and offshore marine farming). Culture-based fisheries should also be given more attention as a viable means of increasing and improving the use of inland lakes, reservoirs and river water resources.

Environmental Management

Improvement of farmer education and awareness, the development and implementation of Codes of Practices and the development of low impact farming systems are subjects that need more attention. In respect of the public perception of aquaculture, community and local authority education on aquaculture issues is needed to be able to provide a balanced picture of the environmental interactions of aquaculture. Environmental impacts on aquaculture are likely to be of increasing concern. Important issues to be addressed include:

* establishment of effective monitoring and management programmes;
* regulation of land-based sources of pollution; and
* protection of aquaculture from the adverse impacts of agricultural, domestic, and industrial activities.

Create an Enabling Policy Environment

Promotion should be made for the further development of government policies that support aquaculture and its development. These should include policy issues relating to the identification and zoning of suitable aquaculture areas and sites, integrated planning (e.g. watershed, lake and coastal planning), and the provision of infrastructure, investment facilities and support activities.

Rather than being actively involved in production, regional governments will increasingly play the role of facilitating development through the creation of an "enabling environment" for sustainable aquaculture.

As pressures increase on resource use, legal and institutional instruments are needed for the recognition of aquaculture as a distinct agricultural sector, which is integrated in resource use and development planning. Legislation is also required for the protection of consumers through improved food safety and quality measures that meet international standards and improve aquaculture management.

Improve the Human Resource

There has been substantial investment in training and education programmes but, if development needs are to be achieved, there is a continuing requirement to improve both the number and quality of human resources. A greater emphasis is required for the improvement of extension support, education and training at all levels, career planning and institutional capacity building. The previous focus on technical training will have to be increased, and emphasis will also need to be given to nontechnical training; such as the development of communication skills. The significant human resource development needs will probably be best met through improved cooperation between the providers of education and training.

Improve the Flow of Knowledge

Activities that improve communication and networking within the region were highlighted as an important means of developing and communicating the knowledge base of aquaculture. Advances in electronic communication, combined with specific cooperative efforts, should be used to increase the information exchange in the region.

Enhance and Maintain Social Harmony

Stakeholder conflict concerning resource use will have to be approached through the better use of conflict management strategies; this should be accompanied by a decentralized and participatory approach to development planning and resource management.

Increase Participation of Farmer Associations

The private sector will continue to play a major role in aquaculture development, but there is a need for better cooperation between aquaculturists to deal with the numerous issues, problems and constraints. The involvement of farmers' associations on a national and regional basis and the encouragement of private-sector participation in aquaculture development will be a key component for success in the coming years.

Move Towards a Market-based Production System

Although the regional and global demand for aquaculture products is expected to increase, it was recognized that there is a need for the sector to move towards a market-oriented approach to the activity, rather than basing development on production. One of the means of achieving this objective is to examine the existing and potential markets for aquaculture products, both within and outside the region, and investigate ways of maximizing intra-regional trade.

Improve Quality and Image

The potential role of HACCP, its application to production systems, the use of drugs and chemicals, appropriate processing and storage technology, and the nature of market/consumer demand are all topics that require investigation and elaboration. Fish handling and postharvest technology could also be improved for the reduction of wastage and the improvement of farm incomes. For each strategic objective, indicative activities have been identified that will require different kinds of support and implementing mechanism to be achieved.

The relevant recommendations of the Conference on Aquaculture in the Third Millennium will be incorporated into the Asian Aquaculture Development Strategy.

Regional and Inter-regional Cooperation

The resources available for designing and implementing aquaculture development projects are no more immune from the current "down-sizing" trends than are those of most other publicly funded development initiatives. There is also an increasing onus on recipients to demonstrate greater accountability for their wise use. The aquaculture sector must now move towards development strategies which favour greater cooperation in collectively addressing the issues important to achieving sustainable development, not just in developing countries, but in all regions of the globe where aquatic animals and plants are farmed. To achieve a level of cooperation which emphasizes complementarity rather than duplication and competition, improved processes need to be identified to partition responsibilities fairly and equitably amongst existing agencies.

Cooperation first requires a focal point to give it life, and then a strategy to sustain it. It is suggested that the most effective focal points for regional development are the Regional Indigenous Organizations (RIOs). They are seen to be particularly important in taking the lead in regional development, since they have been established to serve their constituent member states, and thus have a strong sense of ownership, commitment and responsibility for development in their respective regions. Moreover, they already have a government-mandated framework upon which to structure programmes and projects. They can, therefore, be viewed as being the logical lead agencies around which sustainable development can be pursued. Accordingly then, as the focal point for functional cooperation, they can also be considered for taking on the coordinating role for regional programme participants, which may include: regional intra-governmental representation, universities and research institutes, private-sector interests, industry associations, regional chambers of commerce, nongovernmental organzations (NGOs) and others who have interests in aquaculture development. These same arguments hold true when considering inter-regional cooperation. There is much to be gained through inter-regional cooperation. Specific climatic, cultural or other features usually prevent "wholesale" transplantation of development programmes from one region to another. There are tremendous benefits to be derived from the greater understanding and cooperation made possible by expanding the development process to consider the inter-regional level.

6

Elements of Cooperation

ESTABLISHING DEVELOPMENT PRIORITIES

One of the key elements in measuring how effective we are at achieving cooperation relates to the process of how we go about setting priorities for development. This holds true at all levels: locally, nationally, and both intra-regionally and inter-regionally. With diminished resources available to us, the process through which development priorities are established must necessarily be rigorous and easily defendable from technical and economic, as well as political perspectives. As such, governments and agencies involved in aquaculture development initiatives are being increasingly vigilant to ensure that their decision-making processes satisfy these criteria.

Inevitably, the numbers of worthy projects far outweigh the resources to support them. Clearly, stronger debate and greater wisdom are required to ensure that resources are not applied too thinly nor ineffectively in the hopes of broadening coverage, instead of allocating them to key constraints to development.

Long-term vs. Short-term Goal Setting

Where the process of debate may have once focused on the short-term benefits to be achieved from development projects, in particular tangible assets such as new equipment, buildings and overseas visits to gather information-in effect the narrow view-this track is no longer in keeping with achieving the goals of sustainable development.

In seeking to address the longer-term view, attention needs to be diverted from the more immediate tangible benefits to those achievable within a broader view. This requires a higher level of cooperative thinking and decision making, to ensure that the longer-term goals remain clearly in focus and that scarce internal resources are not side-tracked into achieving short-term objectives which, though they may momentarily appear attractive, are in reality, inconsistent with achieving the major goals.

Internal vs. External Priority Setting

Further, in formulating aquaculture development programmes, greater recognition needs to be given to the importance of ensuring that those priorities which are established are internally driven, and not ones that serve some external agency's agenda, especially if that agenda is not consistent with the recipient's established priorities for development.

This, in itself, is often difficult, since no one wishes to turn down assistance; however, if over the longer term, it means a delay in addressing the important priorities, great care needs to be exercized in proceeding along this path.

Development Agencies-the UN System

On the regional scene, there are a number of levels of cooperation to consider. In Asia for example, there are the various specialized agencies of the UN system which have involvement, at some level, in aquaculture (including inland fisheries) development. These include the Food and Agriculture Organization of the United Nations (FAO), the Asia-Pacific Fishery Commission (FAO/APFIC), the Intergovernmental Oceanographic Commission Sub-Commission for the Western Pacific (IOC/SC-WESTPAC), the United Nations Development Programme (UNDP), the United Nations Environment Programme (UNEP), the United Nations Children's Fund (UNICEF), the United Nations Educational, Scientific and Cultural Organization (UNESCO) and others.

Regional Indigenous Agencies (RIOs)

There are the regional indigenous organizations (RIOs), such as the Association of Southeast Asian Nations (ASEAN), the Network of Aquaculture Centres in Asia-Pacific (NACA), the Southeast Asian Fisheries Development Center (SEAFDEC).and the South Pacific Forum (SPF), which are funded by their member states and, as well, attract programme and project funds from external sources.

Included in the equation are the various country dialogue partners, who provide funding to bilateral and regional development projects; the regional financial institutions; education, training and research institutions; and private-sector interests, all of which are involved in development processes.

Complementing vs. Competing

During the evolution of development agencies in Asia, or in any region, it is to be expected that, amongst them, there are some over-lapping, perhaps even competing interests. To achieve a level of cooperation which emphasizes complementarity rather than duplication and competition in positioning activities on their respective priority agendas, improved processes need to be identified to partition responsibilities fairly and equitably amongst existing agencies.

Areas for Cooperation in Aquaculture Development

When considering topical areas for cooperation in aquaculture development, the following come to mind:

- o Exchange of coastal zone management strategies-how can we achieve sustainable aquaculture development while also protecting the quality of our coastal waters? Are there national or regional strategies which could serve as models for other areas?
- o Development and implementation of aquaculture policies-do all RIOs have an aquaculture development policy that recognizes specific needs and opportunities for aquaculture development? Or are aquaculture development activities included as an appendage of fisheries or agriculture? Are there existing models for national aquaculture policy which could be adapted into a regional format?
- o Marketing has strong proprietary implications for the private sector; nevertheless there is opportunity for generic marketing activities on a regional level that are in support of individual company efforts. Such activities as identification of new export-market opportunities, regional market promotion endeavours and other related activities would be acceptable to most companies.
- o The balance between food security for the domestic market and the need to earn foreign exchange currency through export of higher value products such as prawns and shrimp can be viewed in a regional context.
- o Control of transfer of exotic species-there is an obvious need to minimize the transfer of pathogens and diseases which may accompany movement of

live aquatic animals and their products intra-and inter-regionally. Maintaining biosecurity also needs to be considered in terms of the policies of the World Trade Organization (WTO), as well as in terms of the practicalities and considerable costs associated with establishing and maintaining biosecurity zones.

- While most nations have their own extension programmes developed and implemented internally, the commonalities of aquaculture within regions, and in some cases between regions, would suggest that there are real advantages to focusing on more cooperative approaches. This would apply to the training and education of extension workers, as well as to the delivery of their services.
- For aquaculture chemicals, therapeutants and other treatment products, there is a strong need for regional cooperation in establishing acceptable treatment strategies which meet the needs of the grower, as well as recognize the requirement for environmental sustainability. There needs to be cooperation on which products are acceptable; what limits should be permissible in target and non-target organisms, and in the surrounding environment; how to monitor and enforce compliance; and on how to develop policy and legislation that support the technical requirements. Cooperation involves regional authorities from the exporting states, as well as authorities from the importing states or region. In many cases, there are already existing limits imposed by importing countries. Even so, there is a need for regional cooperative strategies to determine how to meet these limits, especially since rejection or detention from one member state of the region may place

others under suspicion. Representatives of the companies manufacturing the products should also be considered in the equation, since they should be included when it comes to supporting the costs for demonstrating efficacy. Clearly, because of the cost of the equipment and activities involved, it makes more sense to have one regional centre to address these types of problems, rather than leaving it to each country to establish its own centre(s).

- For other areas of research and development, including husbandry and farm systems, education and training, communications and information dissemination, postharvest technology and social dimensions, it has already been demonstrated that regional cooperation offers significant economies of scale.
- When committed to regional cooperation, there are also benefits to be gained in terms of maintaining focus on the political and priority agenda levels. Once agreed to at the regional level, it is more difficult for a member state to unilaterally declare a separate agenda that is inconsistent with the region as a whole.

Inter-regional Cooperation

It is now also clear from previous gatherings, such as the 1997 CIDA Regional Oceans Programmes Workshop (Hinds, 1998), that there is much to be gained through inter-regional cooperation. Specific climatic, cultural or other features usually prevent "wholesale" transplantation of development programmes from one region to another. However, it is acknowledged that there are tremendous benefits to be derived from the greater understanding and cooperation made possible by expanding the development process to consider the inter-regional level.

While geographic distances remain considerable, the advent of the World Wide Web has done much to shrink distances "virtually" and permit opportunity for "south-south" sharing of knowledge, relative to the opportunities and constraints facing people and organizations engaged in aquaculture development in all regions of the globe.

While this particular workshop focused on seven regional projects not specifically devoted to aquaculture, the aim was to arrive at some generic "lessons learned" in regional programme development, and then to examine how these could be applied to the development of subsequent regional projects. This was the first time in which participants from five of CIDA's regional projects were gathered together to formally present their experiences with their projects and how, from a recipient's perspective, the process could be improved. There were 160 participants from over 40 countries-over one-third of whom were women-representing a considerable range of experiences, ideas and approaches.

The five projects represented West Africa (Dioh, 1998), the Caribbean (Saul, 1998), Southeast Asia (Tan, 1998; Jusoh, 1998) and the South Pacific (Maiava, 1998). In addition to the project participants themselves, there were representatives of multilateral and bilateral donor agencies, agencies of the UN, NGOs, the private sector and the universities.

Coming out of this forum was a fresh view of the importance of establishing a strong understanding of how partners develop synergy through cooperation in project design, implementation and measuring progress. The bringing together of the delegates was viewed as the beginning of a process of advancing communications and understanding of how to improve the cooperation required to achieve better designed and delivered projects.

- o With the emphasis on cooperation, most of the lessons learned (Hinds and Bacon, 1998) are directly applicable to sustainable aquaculture development, since they relate to how donors and recipients can interact between and amongst each other to achieve stronger projects. Some of these lessons for the future include: A key ingredient for project success is ensuring input from all stakeholders, particularly the target audience (growers) during all phases of a project, but especially during the design phases. Experience has indicated that when the target audience is excluded from the design process, the project usually encounters considerable difficulty in implementation. When this occurs, there is tremendous loss of development momentum, since the project either needs to be halted for redesign purposes, or in the worst cases, the project fails to achieve its objectives.
- o During project formulation, there needs to be strong regional cooperation in selecting activity centres. There also needs to be regional agreement concerning postproject support for operation of the centres. If regional support is withdrawn at the conclusion of the project, and if the host country alone is left to support the operations, chances for successful continuation of the centre's activities will likely be threatened. Conversely, host countries need to ensure that project centres visibly project a "regional persona" and are not seen as having a national focus.
- o Greater cooperation and coordination need to be exhibited when, as often occurs, several projects are being designed or implemented simultaneously in a region. There is a tendency for projects to

become compartmentalized with linkages restricted to between specific donors and recipients rather than amongst them. In many instances, the opportunities for synergism are overlooked because the coordinating agencies are not fully aware of other projects or are too narrowly focused on the activities of a single project. " While events such as this CIDA workshop are not inexpensive, the argument can be made that periodic gatherings of regional project participants are a wise development strategy. Direct discussions amongst regional project participants are more likely to achieve inter-regional cooperation than if it is imposed by external dialogue partners or agencies.

Cooperation-How do we "Programme" it?

Clearly, cooperation does not happen uni-laterally; it first requires a focal point to give it life, and then a strategy to sustain it. Looking first at the focal point, there are a number of options that could be considered, depending on which agency takes the lead. These could include specific dialogue partners, specialized agencies of the UN, regional financial institutions and perhaps others. However, other than the regional financial institutions, none of these has an exclusive focus on a particular region. As for regional financial institutions, they are lending agencies and their primary focus is not necessarily directed at the detailed strategies involved in sustainable development.

REGIONAL AND INTER-REGIONAL COOPERATION

It is suggested here, that the most effective focal points for regional development are the RIOs. RIOs are seen to be particularly important in taking the lead in regional development since, having been established to serve their constituent member states, they have a strong sense of ownership, commitment and responsibility for development in their respective regions. Moreover, they already have a government-mandated framework upon which to structure programmes and projects. They can, therefore, be viewed as being the logical lead agencies around which sustainable development can be pursued. Accordingly then, as the focal point for functional cooperation, they can also be considered for taking on the coordinating role for regional programme participants, which may include: regional intra-governmental representation, universities and research institutes, private-sector interests, industry associations, regional chambers of commerce, NGOs and others who have interests in aquaculture development.

These same arguments hold true when considering inter-regional cooperation. While it is acknowledged that some RIOs are more developed than others, this only emphasizes the need to include greater inter-RIO dialogue when considering the strategy for achieving sustainability.

Elements of "Sustainability"

Technical Imperatives and Associated Resources

Key elements in the formulation of a sustainable aquaculture development strategy focus on the technical imperatives that need to be addressed, and on marshalling the human and other resources needed to build the elements of the strategy. It is important to recognize that the technical aspects of development cannot be achieved with any lasting effect unless consideration is first given to the institutional architecture needed to support them. All of this needs to be considered within an organizational framework that favours institutional strengthening, capacity building and service delivery with a long-term view. There are plenty of examples of projects that were implemented with too much emphasis placed on the technical contributions of foreign experts, rather than on transferring their expertise and experience to counterparts who have the long-term responsibility for ensuring sustainability. Wherever failure occurred in development projects, it is usually apparent that the technical initiatives were not accompanied by development of the indigenous capacity and capability which would ensure continuation of the project benefits after the departure of foreign experts. The result is that subsequent "trips to the well" were required to 'sustain' the previous project. This clearly is not an acceptable definition of "sustainability".

Institutional Strengthening and Capacity Building

In identifying the key elements of a human resource strategy for achieving sustainable aquaculture development, the focus needs to be on institutional strengthening and capacity building. These are clear prerequisites for better resource management and physically improving the regional development process. Paraphrasing Watson (1998), who discusses institutional strengthening in terms of environmental management, one could suggest the following for sustainable aquaculture development:

- From a project design and implementation perspective, traditional thinking needs to be revised. Where capacity-building projects previously measured achievement of objectives somewhat subjectively, a results-based management (RBM) approach requires that such projects have a strong objective component and that the results achieved are linked to demonstrable improvements in the state of aquaculture development.

- This "new order" implies that project designers must be able to rely on having the appropriate policies in place to support this process of institutional building, and as such, must also have the human intellectual skills available to develop, implement, measure and analyse success towards achieving sustainability.
- Effective policy decisions are the institutional basis for sustainable aquaculture development. As such, when developing policy there is a need for strong communication, cooperation and coordination amongst the RIOs, the education and research institutions and the business community. In assigning the mandate for project development and management to reside with the RIOs, they will then look: i) to the research and education institutions to provide the intellect and innovation needed to improve the technology and methodology, and ii) to the business community to translate it into sustainable aquaculture production.

As a strategy for sustainable capacity building, institutional strengthening requires advisors and counterparts to prepare a comprehensive action plan that is specifically targeted at the resolution of important aquaculture development constraints, or at taking advantage of development opportunities. This process can be considered in two phases:

- Phase I is an outwardly looking "situation assessment" where needs are assessed; priorities for development agreed upon; objectives for capacity building established; and anticipated improvements identified, assuming the objectives are achieved. Strategically, resources need to be targeted at filling institutional gaps rather than being dispersed over a range of secondary activities which cannot be achieved without the critically important institutional strengthening being considered first.

- Phase II is the inwardly looking "baseline assessment" in which institutional constraints facing people and organizations responsible for development are identified. In addressing this phase, the role of the advisors is to assist their counterparts in improving their capabilities through analysis, planning, training and providing opportunities to gain hands-on experience in situations that can be monitored and evaluated objectively using some previously established and accepted performance standards.

A rigorous approach to institutional strengthening is a prerequisite to preventing so called "strategic drift", i.e. fading of original clear-cut objectives and dissipation of targeted resources over a broader range of activities. "Successful implementation of strategy requires that the long-term project objectives are kept in view, activities and outputs are linked to expected outcomes and that frequent journeys down side-tracks without clear objectives, priorities or endpoints are avoided" (Watson 1998).

Framework to Support Technical Activities

In terms of the technical imperatives, "sustainability", in itself, can be viewed from at least three perspectives:

- The first relates to all of the technical components of aquaculture development.
- The technical components must be set in a framework which promotes economical production, considers social obligations to development, and of course, ensures environmental sustainability.
- The last perspective has both an internal and external component; first, aquaculture operations

need to adopt practices that are at least environmentally neutral; and secondly, surrounding industrial and domestic activities must themselves be environmentally neutral to prevent detrimental impacts on nearby farms.

Linkage-Environmental Sustainability and Sustainable Aquaculture Development

Inexorably, sustainability in aquaculture development is linked to environmental sustainability. Much has been studied and written about the need for more vigilance in maintaining an equilibrium between the needs of the shrimp growers and the need to protect coastal mangrove areas as critical nursery habitat for marine species and as an important ecological buffer between tropical marine and terrestrial ecosystems.

This issue, together with others relating to environmental degradation linked directly to aquaculture development, has gained considerable profile and is thus well placed on priority agendas for consideration.

Policy and Legislation Support Technical Initiatives

A second aspect of "sustainability" relates to the policies and legislation that are created to support scientific and technical developments. The establishment of regional environmental, aquatic animal health, transportation and other criteria for aquaculture operations is only truly useful if they are given some effect in law. In formulating sustainable aquaculture strategies, it is important to recognize the downstream legislative process as an integral, though perhaps subsequent, component of the process. The argument can be made that the technical developments may be the easier of the two to accomplish, since legislative development is likely to have many external competing factors that could delay promulgation. This, then, makes a strong case for the focus on RIOs, since they can provide a responsible forum for building credible regional development strategies to promote and protect aquaculture development. RIOs can serve as effective vehicles for consensus building that includes "bottom-up" input at the national level, integrated with an effective "top-down" management function when extended to the regional level.

Sustainability and International Profile

The third aspect of "sustainability" relates to its international profile. As indicated earlier, development funds are limited, and competition for them is vigorous. Most nations within a region share similar problems and opportunities in aquaculture development. The bilateral approach is now viewed as economically inefficient, as well as being too time-consuming if one considers sequential nation-by-nation, project-by-project aquaculture development. Accordingly, cogent arguments can be made for focusing on RIOs which have been assigned strong mandates from their member nations for aquaculture development, and which have equitable arrangements for division of responsibilities among themselves and individual member states, as well as amongst other RIOs within the same region that may have overlapping interests.

As a regional instrument of a group of cooperating governments, a RIO projects a strong image which is the summation of all its member states. Enabled by their charters, RIOs can serve as effective focal points for the development and implementation of regional policy and regulation to support a coordinated approach to sustainable aquaculture development. In doing so, they can also develop, project and sustain a strong regional image of aquaculture as a responsible activity with demonstrated commitment to achieving "best practices", even when some member states, which may be less well developed, are having difficulty in attaining regional performance targets. With the encouragement and support of other members of the RIO, less well-developed states have a greater chance of achieving sustainable development than if they were left to do so on their own.

An example of a successful regionally implemented project focusing on sustainable development is the ASEAN-Canada Cooperative Program On Marine Science (CPMS-II) 1992-1998 (Vigers, pers. comm.). This project also embodies many of the features which have been identified above as being important elements in a successful strategy for regional cooperation. Paraphrasing Vigers (pers. comm.), some of the highlights of the project are as follows:

- The goal of CPMS-II was to support ASEAN's own regional effort to cooperatively optimize its marine resource-based benefits while also maintaining resource integrity and promoting human health protection.
- The broad objective was to upgrade ASEAN marine science capabilities through a programme of training initiatives and through the hands-on execution of three major technical activities:
  - development of tropical marine environmental quality criteria, undertaking pollution monitoring and baseline studies, and
  - investigation of toxic red tides, which cause contamination of shellfish, marine fish kills and human deaths. Participants were from government depart-ments, a university and a research institute representing the Philippines.

Indonesia, Malaysia, Thailand, Singapore and Brunei Darussalam. Project outputs included:

- o through the delivery of more than 100 workshops, technical studies and case studies to over 830 participants, a strengthened ASEAN understanding, technical knowledge and institutional capacity in marine environmental management leading to greater recognition of the need to work regionally;
- o strengthening of human resource skills, knowledge and technical capabilities through the placement of graduate student trainees, short-term training attachments and study tours;
- o establishment of a cadre of ASEAN specialists, and the development of a networking system amongst them;
- o formulation of a common set of ASEAN marine environmental quality criteria for key environmental parameters;development of toxicity testing and analytical protocols that are contributing to data comparability and reliability within the region;
- o regional agreement on standard methods, procedures and quality analysis/quality control (QA/QC) methods for sampling and analyses;
- o identification of regional marine environmental quality targets and establishment of regional baseline data sets;
- o establishment of a red tide surveillance and advisory network;
- o an increasingly harmonized approach to regulatory implementation monitoring, enforcement and policy development in marine environmental management that recognizes the conflicting needs of

industrial development, population and economic growth, as well as ecological sustainability; and

- o closer technical cooperation between Canada and ASEAN.

One of the key features of this project, which has particular relevance to sustainable aquaculture development, concerns the management strategy for the project. Specifically, the Canadian Executing Agency (CEA) adopted an advisory role to the Project Steering Committee and Technical Working Groups. The CEA, working in a consultative manner, administered funds on behalf of CIDA, sourced technical specialists to be made available to the project, maintained overall project momentum, and provided an assistant project coordinator to help in managing and administering the many project activities occurring in the region from the Project Execution Centre located within the Department of Fisheries, Government of Malaysia, Kuala Lumpur.

However, the Project Coordinator was selected by ASEAN, and the direction of the project and the forces impacting on its focus were very definitely ASEAN. This was in keeping with the project design philosophy of having ASEAN assume the lead role when it came to responsibility for the conduct of the project. This approach was possible because ASEAN, as a well-structured RIO, has the strength needed to assume this level of responsibility. As it turns out, this approach was successful. Moreover, when analysing the inputs to human resource development and physical resources dedicated to the project, it was evident that for every two dollars Canada contributed to the project, ASEAN contributed three dollars.

A Regional Sustainable Aquaculture Development Strategy for Consideration: It is fitting, though not coincidental, that we are discussing models for regional sustainable aquaculture development, here in Asia. After all, it is generally agreed that aquaculture had its beginnings in Asia, probably at least two millennia ago. Moreover, in terms of RIOs, it is generally agreed that Asian organizations have probably achieved a higher level of development to date, than those in other parts of the globe. Accordingly, looking eventually at inter-regional strategies as a means of promoting greater sustainability in aquaculture development, it makes sense to first ensure that there is a solid model in at least one region upon which to build future initiatives.

Synthesising the elements from the foregoing discussion, we would suggest that the following are the major considerations in designing a regional strategy for sustainable aquaculture development: A RIO as Lead Agency: For all of the reasons indicated earlier in this discussion, a suitable RIO to take the lead in the development process is a primary consideration. In Asia, there are currently three RIOs (NACA, SEAFDEC, ASEAN) with strong ties to aquaculture development. Of these, NACA with its regional network of centres, is seen as being the focal point for a strong regional approach to aquaculture development. There are several reasons to support this suggestion, including NACA's:

- o exclusive focus on aquaculture development through regional cooperation;
- o approximately 20 member and participating nations encompass most of the people of Asia;
- o ability to provide the "bottom-up" elements from an R&D perspective;
- o endorsement by FAO and the confidence of other agencies; and
- o experience in designing and delivering effective R&D programmes.

It makes sense then, that with this profile and level of regional representation, favourable consideration needs to be given to NACA as a lead agency.

However, consideration needs also be given to the other two RIOs. SEAFDEC has a well-developed aquaculture programme and a strong history of institutional strengthening in partnership with the governments of its member states. Moreover, SEAFDEC also has considerable experience in technology transfer initiatives. Specifically, many of SEAFDEC's pilot-scale projects have involved bridging the gap between the R&D institution laboratory bench level, and commercialization through an industry partner. As a facilitator of aquaculture industrial development, SEAFDEC too, clearly has an important role to play as a lead agency for development.

ASEAN is a formal political organization representing governments of seven countries, has close ties with other countries in the region, and has well-established connections to multilateral and other development organizations. As such, ASEAN will be important in providing the political leadership to keep aquaculture development high on the priority agenda. To subsequently enable specific development initiatives, ASEAN will also be important in ensuring that funds are sourced, either internally or in partnership with other NACA members, or in broader partnerships involving bilateral and multilateral development agencies.

It is concluded then, that while all three RIOs are currently pursuing aquaculture development in Asia, that a more synergistic approach focused on complementing each other's strengths could be achieved through a partnership involving the technical and institutional strengths of NACA and SEAFDEC, combined with the political expertise resident within ASEAN. One could envisage the aquaculture development initiatives undertaken by these three organizations being integrated and consolidated in a manner which would give a more effective regional voice to sustainable aquaculture development issues.

It is suggested that initially there needs to be a forum in which senior representatives of NACA, SEAFDEC and ASEAN can meet to explore possibilities and hopefully develop a common front on approaches to sustainable aquaculture development in Asia. Also included in the forum could be traditional dialogue partners, FAO and other specialized agencies who would be viewed as critical to the process of establishing the priority agendas for regional development. Such a high profile forum, aimed at consolidating regional efforts in aquaculture development, should also emit positive signals to attract regional and global support for initiatives emanating from this cooperative approach.

Assuming that an agreement can be reached, the next step could be a five-year pilot programme designed to permit the integration and consolidation process to proceed on an experimental basis. This would provide an adequate timeframe in which to work through the various issues associated with the move in this direction.

The intention here is that NACA, SEAFDEC and ASEAN would carry on their normal work programmes, but would also focus on determining how they can optimize combined expertise to accelerate development, as well as reorganize .activities amongst themselves to avoid duplication of effort. For example, there could be "trade-offs" in the form of one agency taking the lead for a specific technical initiative, with others adopting support roles. This would not only reduce duplication in the system, but could enhance complementary activities, in place of competition.

Setting Priority Agendas

Placing aquaculture high on the priority list of regional development initiatives will be a rigorous exercise in consensus building. Nevertheless, the proposed NACA/SEAFDEC partnership with support from ASEAN would have considerable experience and expertise in this regard. The principal task will be designing and implementing the process by which development initiatives can be moved from scientific and technical design, through the bureaucratic level, to the political level where they can garner greater profile and compete more effectively with other issues on the political agenda. This proposed strategy, which also provides for appropriate monitoring and evaluation, is illustrated in Figure 1. It is envisaged that the chain of events in the process may resemble the following: The private sector identifies the need for a particular research and development initiative, which has generic implications for the industry as a whole, or for some subsector of it. For example, a particular disease issue in a farmed species may require better diagnostic and treatment strategies, or some other issue that, when addressed adequately, can improve the standard of aquaculture in general. The need is communicated to the RIOs.

- NACA/SEAFDEC, with their expertise in undertaking R&D programmes, identify the technical elements required to address the issues; compare these to the inventories of specialists within their member centres, and prepare the project design. If external expertise is required, this would be identified at this stage. A commercialization strategy would also be included, using private-sector participation to ensure that results from the R&D component are translated into commercial benefits.

- SEAFDEC, with its expertise in the commercialization of R&D initiatives, would play a lead role in this aspect of the programme.

  ASEAN's political expertise would be enlisted to move the technical plan forward for consideration by the appropriate bureaucratic levels (usually by economic planning units of representative regional governments).
- When senior bureaucratic approval is received, ASEAN would be enlisted to move the process to the political level for approval. In doing so, consideration would also be given to strategies for publicising the initiative in a way which is likely to favour a positive political decision. For instance, assuming that there are to be spin-off benefits to the aquaculture service sector and others, the support of regional chambers of commerce could be enlisted to help promote or endorse the agenda.
- Following approvals, project implementation of the technical components through NACA/SEAFDEC, and then the commercialization components through NACA/SEAFDEC and the private-sector partner(s) would proceed.
- At appropriate points throughout the project, ASEAN coordinates internal and external monitoring and evaluation activities.
- It is expected that the project, when completed, will demonstrate a measurable advance in development, and address the specific needs identified at the start.
- The model is cyclical in nature and thus provides for sequences of project initiatives all directed at moving sustainable development forward.

In moving the priority "through the system" in this manner, there is good opportunity for integrating the bottom-up concerns, prominent at the local and national levels, with a more benevolent top-down management approach necessary for a successful regional approach to cooperation.

Institutional Strengthening and Capacity Building: As indicated earlier in this paper, successful programmes have focused on these elements as a means of ensuring sustainability. While most human resource development has focused on providing graduate and postgraduate training to incumbents in specific programmes, consideration should also be given to more formal university and technical institutional curricula focusing on aquaculture development. If there is stronger formalized grounding provided in the fundamentals of aquaculture development, counterparts would be better prepared, and existing operations would be better able to absorb new project activities into their line operations.

While there is a need to continue to develop the academic ranks of aquaculture specialists who will undertake R&D projects in universities, research institutes and government laboratories, there is also a need to focus on the training required for aquaculturists whose career path will be towards operations and management of commercial ventures. Accordingly, while their curriculum will include many of the same subjects as those of the scientific and technical specialists, they would also focus attention on the practical aspects of aquaculture operations, including resource engineering, geotechnical topics, environmental sciences, aquatic animal health and nutrition management, husbandry, and especially business management.

The latter is to ensure that they develop the capacity to evaluate and integrate attractive technical innovations into a company's operations, within the limits allowed by the company's financial strength.

Extending the Model Regionally and Inter-regionally: NACA's member and participating states include several nations outside Southeast Asia. Thus, while the first step in building a stronger regional approach to sustainable aquaculture development may be on the "sub-region" of Southeast Asia, the ultimate intent is that the structural elements of the new partnership can be extended to other states in the region, through NACA's central coordinating role.

Following the development of this "Asia model", it is conceivable that a forum could be held in which RIOs from other areas of the globe would gather to consider how this particular model could be shaped to address other regions' needs for sustainable aquaculture development. Here, it is envisaged that the process utilized is as important as the results achieved. Specifically, assuming that such a model comes to life, it would be particularly useful for other RIOs to be able to share the experiences encountered as it develops. This would allow them to develop some context of how analogous models could be developed in their own regions.

Even better than waiting for successful development of the currently proposed model, would be to share the experiences inter-regionally, as they are encountered. This would provide for more inter-regional dialogue, more rapid dissemination of useful information, and serve as a catalyst for the overall thrust of greater inter-regional cooperation in aquaculture development.

Ultimately, a more efficient and productive approach to cooperation in sustainable aquaculture development will translate into visibly positive actions with respect to addressing the existing global protein deficit problem.

7

Current Status of Aquaculture

Compared with fishing, aquaculture is currently of little commercial significance to the Pacific Islands, with one important exception, black pearl farming, which is virtually confined to eastern Polynesia. Elsewhere in the Pacific, considerable development is needed before aquaculture can be considered economically sustainable. Shrimp (Penaeus spp.) farming has been a focus of commercial development in several islands with varying degrees of success; tilapia (Oreochromis niloticus) aquaculture has entered the subsistence economy in some areas, and seaweed (Kappaphycus spp.) is a future commercial export prospect. The culture of other marine and freshwater species is, however, generally still at the experimental or "backyard" stage.

The expansion of aquaculture in the Pacific will depend on providing better production methods for species currently being farmed, and techniques for propagating and growing the "new" species described above. These methods and techniques should be simple and flexible so that they can be adapted to the context of the Pacific Islands environment and to the market constraints (local and export markets). This approach should favour systems integrating fisheries and mariculture with low investment and operating costs and simple technical production processes. This should be done in association with pilot commercial-scale operations to test and demonstrate the economic viability of the methods proposed. This will require research combined with assistance, training and education programmes.

Pacific Island nations have many attributes that favour development of aquaculture and stock enhancement in the coastal zone. These are as follows: a great diversity of coral reef species which are in high demand, proximity to major aquaculture and seafood markets in Asia, availability of suitable growout sites in pristine habitats, geographic conditions which favour restocking and stock enhancement, a relatively inexpensive labour force, and a tradition of working with marine resources. Although these confer many advantages on the region in terms of aquaculture development and stock enhancement, there are still several constraints for such enterprises in the Pacific, which include limited domestic markets, high added-value export markets targeted, transport problems, socio-economic factors, fragile habitats, limited fresh water, and cyclones. Some of the best opportunities for aquaculture development in the Pacific are in the aquarium trade (coral reef fish, hard corals, soft corals), the live seafood markets (e.g. groupers, spiny lobsters, abalone, crabs) and the pharmaceutical industry (e.g. algae, sponges, soft corals). In all cases, the products are of high value and can be grown in small areas with relatively simple technology.

This review covers the insular Pacific as defined by the work-area of the Secretariat of the Pacific Community (SPC).

Compared to fishing, aquaculture is currently of little commercial significance in the insular Pacific, with one important exception-black pearl farming-and this is virtually confined to eastern Polynesia. Elsewhere in the Pacific, considerable development is needed before aquaculture can be considered economically sustainable. Shrimp (Penaeus spp.) farming has been a focus of commercial development in several islands over the past 30 years, with varying degrees of success; tilapia (Oreochromis niloticus) aquaculture has entered the subsistence economy in some areas, and seaweed (Kappaphyceus spp.) is considered a future commercial export prospect by the region. The culture of other marine and freshwater species is generally, however, still at the experimental or "backyard" stage.

Aquaculture is a relatively new development in the region, and in most Pacific Islands where it has been attempted; its history goes back less than 30 years. There is no fund of traditional knowledge for culturing fish and shellfish, just catching them, except in very specialized instances and areas. There is thus no great resource of aquacultural skill or infrastructure. This steep development path has perhaps not been taken into account in some development projects, which have often had unrealistic short-term aims and lacked follow-up.

Despite the comparatively minor penetration of aquaculture into Pacific Island economies, and despite the loss of interest by most of the international development community after many short-term project failures, several Pacific Island governments have accepted the challenge. They recognize that expansion in capture fisheries is limited, and have made substantial investments in freshwater aquaculture and mariculture, often in concert with external sources of development assistance.

Pacific Islanders cannot turn away from the sea. It is the greatest resource they have.

The Potential Importance of Aquaculture to Pacific Islands: A major problem facing most of the island nations in the Pacific is that they have relatively few opportunities to generate income (Adams, 1998). The economies of most Pacific countries are limited due to small landmasses, few terrestrial resources and low numbers of inhabitants. To ensure further development, island nations must make the most of the one important resource they all have-the sea. Through the joint efforts of the Forum Fisheries Agency (FFA) and the SPC Oceanic Fisheries Programme, island nations are deriving major inputs to their economies by fishing for tuna, or by selling access rights to tuna, within the large maritime zones under their control. However, valuable, sustainable harvests are also possible from inshore waters and coral reefs.

The potential for increased well-being of coastal communities from the responsibleuse of their inshore marine resources arises because the inshore habitats surrounding Pacific nations support a great diversity of economically important species (Wright and Hill, 1993; Dalzell et al., 1996; Bell and Gervis, 1999). Traditionally, these animals were harvested at subsistence levels. More recently, however, development of export markets has provided coastal communities with opportunities to earn income from the inshore fisheries species. Unfortunately, the transition from a subsistence to a market economy has usually been far from ideal: chronic overfishing has occurred in some areas. In such places, there are now too few of the prized animals to sustain reasonable harvests. Destructive fishing methods have compounded the problem by degrading some habitats to the point where they cannot support the valuable species (McManus, 1997). Pacific Island countries now recognize that aquaculture provides one of the few long-term, sustainable, ways of deriving benefits from inshore fisheries resources

(Williams, 1996). This view of aquaculture as a priority area for continued sustainable development was reinforced as part of a consensus member country statement arising from the 2nd SPC Fisheries Management Workshop in 1998:

> "*...Regional fisheries managers have focussed on establishing regimes to sustain inshore fisheries. This is supported by a strategy to divert demand and fishing pressure to alternative activities, mostly to offshore fishing and into aquaculture.*

Development of fisheries management varies from country to country reflecting the differing stages of economic development and levels of need. In some countries the need is to encourage economic activities and to generate income for rural villages; in other countries the need is to restrict or limit fishing. Yet in other countries the need to involve all stakeholders in the management system has evolved into community-based decision-making and control. There is now broad acceptance that marine resources cannot be managed in isolation from other users, or by one government agency so that an integrated and co-ordinated approach should be taken. In many circumstances, because of the smallness of the islands, an island system-management approach is the desirable option.

Aquaculture, as an alternative activity, is still at a preliminary stage of economic development in most PIC3, but is of enormous future significance. For aquaculture to realise its full potential to the economies of PIC in a sustainable way will require a considerable degree of international support. PIC have endorsed a strategy to harness and prioritise such support at the regional institutional level. Several PIC already devote significant national resources to this subsector and this trend will continue as benefits are realised..."

Advantages of Pacific Islands for Aquaculture: Pacific Island nations have many attributes that favour development of aquaculture and stock enhancement in the coastal zone. These include:

- o A great diversity of species associated with coral reefs that are in high demand for:
- o the aquaculture and seafood markets in Asia (e.g. napoleon wrasse, groupers, sea cucumbers, spiny lobsters, trochus, pearl oysters, giant clams, green snail);
- o the marine aquarium trade (e.g. clownfish, angelfish, hard corals, soft corals, giant clams); and
- o the pharmaceutical trade (e.g. algae, sponges, soft corals, sea horses).
- o Proximity to the major aquaculture and seafood markets of Asia-flight times are short enough to ensure that many species can be shipped alive to Asia.
- o Availability of suitable growout sites in pristine habitats-coral reef lagoons create the calm conditions essential for culture of many species. The favourable environmental conditions should be the opportunity to develop green label products to get better prices on the international market.
- o Geography that favours restocking and stock enhancement-most Pacific countries are small islands, or groups of small islands, surrounded by deep water. Cultured juveniles released into the inshore waters of island ecosystems cannot emigrate, and are therefore relatively easy to recapture.
- o A relatively inexpensive labour force-expectations for financial return on labour are low in many Pacific countries relative to developed countries.

- o A tradition of working with marine resources-coastal communities are already familiar with the basic biology of many species.

Constraints to Aquaculture in Pacific Islands

Although the attributes listed above confer many advantages on the region for development of aquaculture and stock enhancement, there are also several constraints to such enterprises in the Pacific. Many of these have been identified previously by Uwate and Kunatuba (1983), Munro (1993) and Bell and Gervis (1999). They include:

- o Limited domestic markets. Local markets for the fresh products of aquaculture in the Pacific are small, and with the exception of very limited opportunities in the restaurant trade, usually offer low prices. Thus any large-scale aquaculture development in the Pacific catering to the trade in seafood will depend heavily on export markets.
- o High added-value export markets targeted. These are most often regional and fluctuating markets (e.g. live reef fish). Thus any aquaculture development will depend on the capacity to follow the market trends, and to fulfil the demand in time.
- o Transport problems. The high cost of shipping in the Pacific adds considerably to the cost of producing and exporting aquaculture products. Poor internal transport services restrict opportunities to grow perishable products in remote locations, and limited international air connections inhibit continuity of supply to export markets. Transport arrangements dictate that species cultured for export need to be of high value and low weight. Alternatively, the products must be non-perishable,

e.g. bêche-de-mer (processed sea cucumbers) or frozen, so that they can be shipped by sea.

- o Social and economic factors. Many of the smaller island nations lack the infrastructure, capital and skilled labour required to implement aquaculture, particularly where hatcheries are involved. Sustained assistance from developed countries is needed to implement and operate stock enhancement programmes until they become self-funding (Bell, 1999a). The traditional marine tenure systems in place in many countries (Ruddle et al., 1992) also add complexity to the process of negotiating access and tenure to sites for aquaculture.
- o Fragile habitats. Coral reef ecosystems on many of the smaller island nations have evolved in a nutrient-poor environment. Additions of nutrients, e.g.through uneaten and undigested formulated diets for carnivorous fish in cage culture (Beveridge, 1987; Landesman, 1995; Stewart, 1997), can be expected to change the ecosystem in favour of algae and herbivores. Such changes are likely to be unacceptable, particularly to the tourist industry. This constraint is particularly relevant to lagoonal habitats, but would not apply to locations that have good flushing to the open ocean.
- o Freshwater is limited, except for the large islands of Melanesia, which have extensive river systems. Prospects for freshwater aquaculture are thus limited. Even in areas with significant fluvial development, the indigenous freshwater ichthyofauna is generally unfavourable for economic culture and species for freshwater aquaculture have been imported.

- o Cyclones. Countries in the cyclone belt can expect aquaculture installations to be damaged intermittently by large swells and strong winds.

CURRENT STATUS OF AQUACULTURE

Aquaculture Systems: Profitable aquaculture of penaeid shrimps and blacklip pearl oysters has now been established in some areas of the Pacific by commercial interests. Stand-alone enterprises producing penaeid shrimps for export markets are firmly established in New Caledonia and Fiji and were so in Solomon Islands until recently. These enterprises are applying technology developed originally in Japan, Taiwan Province of China and France, and now common place throughout the tropics.

A large, sustainable, industry for culturing pearls using the blacklip pearl oyster (Pinctada margaritifera) has been established in the Tuamotu Archipelago, French Polynesia, and on a couple of atolls in the Cook Islands (Fassler, 1995). In French Polynesia, the value of cultured pearls exceeds US$150 million per year. In Cook Islands, the industry is currently worth US$5 million and is the second largest source of revenue for the country after tourism.

Black pearl farming in French Polynesia and Cook Islands, and shrimp aquaculture in New Caledonia represented more than 98 percent of the total value of aquaculture production estimated in 1996.

Blacklip Pearl Oysters: Small-scale culture of pearl oysters is under way in Fiji, Marshall Islands, the Federated States of Micronesia, Solomon Islands, Kiribati and Tonga. In some places, e.g. Kiribati, development is based on spat produced in hatcheries, whereas in others, e.g. Solomon Islands, development is geared towards finding ways that coastal villagers can catch and grow wild spat (Friedman et al., 1998). Current research is concentrating on assessing the economic viability of pearl farming in Solomon Islands, Fiji and Kiribati, and comparing growth, survival and pearl quality of oysters derived from wild and hatchery-reared spat.

Giant Clams: Small-scale enterprises in Solomon Islands, Palau, Marshall Islands, Cook Islands, Tonga and American Samoa supply five species of giant clams (Tridacna crocea, T. derasa, T. gigas, T. maxima and T. squamosa) to the marine aquarium trade (Foyle et al., 1997; Hart et al., 1998). Production of giant clams for enhancement of wild stocks is also under way in Solomon Islands, Fiji, Cook Islands and Western Samoa (Bell et al., 1997a; Bell, 1999b). Several of these countries also have the capacity to produce giant clams for the sashimi market in Okinawa, and as a live product for markets in China, Hong Kong Special Administrative Region and Taiwan Province of China (Bell et al., 1997b).

Sea Cucumbers: Research has commenced to assess the viability of producing sea cucumbers in hatcheries for enhancement of wild stocks. There are three steps in this process: developing methods for cost-effective mass production of juveniles, learning to release the cultured juveniles in ways that maximize their survival, and evaluating the economic impact of releasing cultured juveniles into existing fisheries. Currently, the focus is on development of methods for the mass rearing of Holothuria scabra, H. fuscogilva and Actinopyga mauritiana, three of the most valuable sea cucumbers in the region. To date, ICLARM has demonstrated that H. scabra is relatively easy and cheap to rear (Battaglene and Bell, 1999), and that A. mauritiana grows relatively rapidly at high densities (Ramofafia et al., 1997). Initial research on H. scabra indicates that this species has much potential for stock enhancement.

Other Species: Technology for propagating and releasing cultured juveniles of green snail and trochus has been transferred to the Pacific through projects in Tonga (JICA) and Vanuatu (Austrailian Centre for International Agricultural Research, ACIAR). The Overseas Fishery Cooperation Foundation (OFCF) is also implementing a stock enhancement programme for both species in Solomon Islands. Production of the marine alga Kappaphycus is well established by coastal villagers in Kiribati and Fiji, and sponges are being cultured in the Federated States of Micronesia. Milkfish are being cultured as live bait for the tuna industry in Guam, and there is considerable interest in this activity by several other countries.

Policy and Institutional Framework: The independent Pacific Islands (Cook Islands, Federated States of Micronesia, Fiji, Kiribati, Marshall Islands, Nauru, Niue, Palau, Papua New Guinea, Samoa, Solomon Islands, Tonga, Tuvalu, Vanuatu) generally lack any specific provision for aquaculture in the legislation, but many include a statement about aquaculture in national development plans.

Whilst prospects for inland aquaculture are limited by geography, the custom of communal tenure of coastal marine areas may be incongruous with private-sector farm ownership in many areas, unless the development is carefully managed. Tonga is currently developing a comprehensive legislative basis for future aquaculture development, whilst the Cook Islands has adapted traditional systems into a legal basis for pearl farm management. Other countries are taking a more ad hoc approach and trying to adapt land-based systems of state leases for mariculture development, or encouraging development only by traditional reef custodians.

Fiji: Aquaculture was first started in 1953, when tilapia were introduced as a protein source for pig farming. The first directed efforts occurred after the United States Peace Corps and a JICA project assisted the Fiji government in developing freshwater aquaculture methods in the 1980s. Varied success was achieved with shrimp, Kappaphycus, oyster, mussel, Macrobrachium, carp and tilapia, but now major investment is promoted by the government through a Commodity Development Fund. Currently, there are three shrimp farms and hatchery; 20 milkfish ponds for longline bait; and one industrial, seven commercial and 215 subsistence farms and six hatcheries that produced 243 mt of tilapia in 1998. Eucheuma cottonii is cultured by a total of 182 farms, producing an estimated 1 500 mt in 1999. Experimental pearl farming is conducted, and there is one commercial farm that has been operating for two decades. There is also a giant clam hatchery, the Naduruloulou Government Freshwater Aquaculture Research Station, the Makogai Government Mariculture Research Station, and the University of the South Pacific (USP) Aquaculture Teaching-School Teaching Ponds (E. Ledua, 1999, unpublished data).

Guam: There was an increasing trend in annual aquaculture production between 1990 and 1995 (Clarke, 1997). In 1995, aquaculture production was 205 mt valued at US$1.6, derived mainly from the farming of tilapia, marine shrimps, Chinese catfish and milkfish. The potential for the development of other species (giant clams, top shell, striped mullet and groupers) is being explored by Guam-based research facilities.

Kiribati: Eighty hectares of mikfish ponds, originally set up to provide livebait for tuna pole-and-lining, have been operating for several decades. Kappaphycus alvarezii has been cultured in the Phoenix, Line and Gilbert groups for 15 years. There is an OFCF project on sea cucumber rearing and an ACIAR project on pearl farming (T. Tekinaiti, 1999, unpublished data).

Nauru: Milkfish been farmed for at least one century, but competition from introduced tilapia (O. mossambicus) occurred in Buada Lagoon and culture lapsed. It has recently been revived, with O. niloticus and milkfish being raised in 11 ponds (F. Alefaio, 1999, unpublished data).

Northern Marianas: Aquaculture is limited to raising penaeid shrimp and tilapia (Clarke, 1997). Annual production is estimated at 1 250 kg for marine shrimp, 200 kg for freshwater prawns and 4 200 kg for tilapia. There are fewer than three commercial farms. The total value of products was estimated as US$25 000 in 1996.

Nouvelle Caledonie: Shrimp production achieved 1 500 mt for the first time in 1998 (Labrosse et al., in press), after disease problems which occurred in the early 1990s. The value of marine shrimp production is about US$10 million. About 45 mt of oysters (Crassostrea gigas) are produced for the local market. Experimental culture of local oyster and giant clam is also in progress (Etaix-Bonnin, 1999).

Palau: The Micronesian Mariculture Demonstration Centre pioneered giant clam culture, as well as trochus and soft corals.

Papua New Guinea: Aquaculture started 40 years ago, with several aquaculture stations along the coast and highlands to encourage subsistence culture, mainly of Cyprinus carpio. There are 300 carp farms in operation. Trout were introduced in the 1940s, and the Kotuni Trout Farm was in operation from 1973-1984. There are three newer farms, but only two are currently operating, with a production of 15 mt. A hatchery was started in 1996.

Culture of barramundi (Lates calcarifer) has been started in Madang on the site of a failed 9 ha pond originally used for redclaw (Cherax quadricarinatus), but production has not yet been established (J. Wani, 1999, unpublished data).

Samoa: Farming trials were conducted on tilapia in 1954 by the SPC and on Macrobrachium in 1971 by the FAO, but were not successful. Seaweed, giant clam, green mussel and redclaw farming were also tried. A new national economic strategy promotes aquaculture, and this is being actively developed by the current AusAID village fisheries extension project using tilapia, mullet and giant clam (A. Mulipola, 1999, unpublished data).

Solomon Islands: There are two shrimp farms and several village-based enterprises rearing giant clams and hard corals for the aquarium trade and a demonstration black pearl farm in the Western Province. The ICLARM Coastal Aquaculture Centre has also developed methods for the propagation of sandfish (Holothuria scabra) and the capture and culture of wild postlarval coral reef fish for the aquarium trade. Between 1996 and 1999 there was an OFCF project on green snail & trochus (E. Oreihaka, 1999, unpublished data), but the future of the aquaculture-oriented Institute of Marine Resources of the University of the South Pacific is uncertain.

Prospects for Further Development

Some of the best opportunities for development of aquaculture in the Pacific are in the aquarium trade and live seafood markets (e.g. napoleon wrasse, groupers, sea cucumbers, spiny lobsters, trochus, pearl oysters, giant clams, green snail, abalone, crabs, clownfish, angelfish, hard corals, soft corals) and the pharmaceutical industry (e.g. algae, sponges, soft corals, sea horses) (Bell and Gervis, 1999; Bell, 1999c). In all cases, the products are of high value and can be grown in small areas with relatively simple technology.

Initiatives by FAO, the World Fish Center (ICLARM), the Center for Tropical and Subtropical Aquaculture (CTSA) and bilateral donors have concentrated on establishing the culture of pearl oysters outside eastern Polynesia, developing small-scale aquaculture enterprises for other species, and providing basic training in aquaculture and stock enhancement to fisheries staff in several of the countries.

The proceedings of the workshop entitled Present and Future of Aquaculture in the Pacific organized jointly by the South Pacific Aquaculture Development Project (SPADP) and the Japanese International Cooperation Agency (JICA) in Tonga in November, 1995 (Anon., 1996) presented the status of aquaculture in the region.

The continued expansion of aquaculture in the Pacific will depend on providing better methods of production for species currently under cultivation, and techniques for propagating and growing the "new" species described above (Bell, 1999c). These methods and techniques should be simple and flexible enough to be easily adapted to the context of the Pacific Islands environment and to the constraints of local and export markets. This approach should promote systems integrating fisheries and mariculture, with low cost methods of production). This should be associated with pilot commercial scale operations to test and demonstrate the economic viability of the methods proposed. This will need research coupled with assistance, training and education programmes.

In recognition of these needs, three organizations (SPC, ICLARM and USP) have recently joined forces to produce a "Regional Strategy for the Development of Aquaculture". Under this strategy, SPC will be the regional focal point for aquaculture and willconvene regular meetings of island nations to identify needs, determine priorities and put organisations and individuals in touch with each other. ICLARM will undertake long-term research to devise and test economically and environmentally sustainable methods for restocking, stock enhancement and farming; and USP will developdegree and diploma course components and aquaculture vocational training, and contribute to research through higher degree programmes. The other functions necessary for the expansion of sustainable aquaculture in the region, e.g. marketing, legislation, environmental protection and quarantine, will be integrated progressively through association with the Strategy.

This Pacific Islands Regional Aquaculture Strategy represents an opportunity to reinforce inter-regional cooperation based on research, training and information exchanges (including cooperation through NACA with Southeast Asian countries). It may promote more investment from Asia and better conditions for access to Asian markets.

FISH, THEIR "FRIENDS," AND THE FREE MARKET

Introduction

Although marine fish, as residents of interconnected bodies of water that span the globe, are continually mobile and nearly impossible to localize or claim, as a resource for human consumption they have become a source of political debate and conflict between very localized states and groups of people. Currently, as the threatened state of global fisheries resources becomes increasingly prominent, the lines for disagreement, conflict, and negotiation are diverse and run the spectrum from international organizations to groups of different economic and social strata at the level of coastal communities. To people around the world of varying wealth, nationality, economic system, culture, and position of power, fish play a crucial role in their tastes, traditions, and needs for survival. This, more than what needs to be done ecologically to restore the world's fish resources, is what makes the situation so complex.

One effort to simultaneously resolve the seemingly disparate elements of the problem-the economic, the environmental, and the developmental-is the concept of sustainable development. In the context of fisheries, this concept entails being able to manage the pressures for the production and trade of fish, the strain this places on the resource base, and the needs of those people that depend on the resource for livelihood and food security to develop in an equitable manner. As ebbs or flows in one sphere impact the security of the others, solutions in no one sphere can be seen as representing a sustainable solution if the other two are neglected.

Of so much importance to so many people around the world, individuals, organizations, and states are searching for ways to alleviate the intense pressures on fisheries resources without completely sacrificing the interests of humans in using the resource for consumption and a source of livelihood. One current example of this is the proposal among states and international conservation and sustainable development organizations to use subsidy reform in the World Trade Organization (WTO) to pursue sustainable development objectives. While subsidies are a crucial contributor to the overfishing that occurs in many of the world's most important fisheries, the questions is whether or not such a reform can deliver on all three pillars of sustainable development. Subsidy elimination or reform in the fisheries sector through the WTO holds several legitimate possibilities for better resource use and alleviation of perverse economic incentives in fisheries products production that suggest advancement in important aspects of human development in developing coastal countries. This paper intends to explain, though, the limitations of this action and the possible threats to development and resource use in developing countries posed by such an endorsement of free market principles, especially given certain intranational realities of many of the most fish-wealthy developing countries.

Background on the Current State of Depletion in Fisheries Resources

The first step towards an accurate conception of current global fish-use, and possibly the least debatable aspect of the issue, is recognition of the precarious biological and environmental situation of the world's fish stocks in the face of increased global consumption of fish in the past 30 years. Since 1973, global consumption of fish has doubled. Growth in demand for fish as a source of protein in the developing world has accounted for ninety percent of this growth, while developed countries' share in world fish consumption has decreased. Fish is consumed most highly by far in Asia, and next highly in Europe. Fish consumption is generally higher in developed countries, but may play a larger role in food security in developing countries. At the same time, populations of fish in almost every part of the globe have declined to levels well below those they have historically and naturally held. The World Wide Fund for Nature (WWF) reports that 73 to 75 percent of the world's major fisheries are overexploited, fully exploited, or recovering from depletion. An article from the United Nations Food and Agriculture Organization (FAO) reports that, "Judging from known fish stocks and resources of traditional fisheries, the total marine catches from most of the main fishing areas in the Atlantic Ocean and some in the Pacific Ocean would appear to have reached their maximum potential some years ago." Global demand remains high, and the world's fisheries seem unable to support it in any sustainable way.

The link between increased global demand for fisheries products and the current level of stock exploitation is the dramatic increase in production that has occurred on an international level in the same time period. While 44 million tons of fish were captured for food in 1973, 65 were in 1997. Since the 1980s, developing countries have taken the lead in wild fish capture and now account for over 70 percent of global fish production. After such high levels in the last 30 years, though, wild fish production has stagnated; the state of exploitation in main fishing areas has actually remained more or less unchanged since the early 1990s. The world's fish stocks cannot continue to sustain such intensive levels of capture, a reality which has serious implications for both those on the production and consumption side of fishing.

The international trade of fish is also an important dynamic in the current situation. About 40 percent of world output by value of fish was traded across international borders in 1998. Since the early 1970s, the direction of fish trade has experienced an "about-face:" developed countries were net exporters of 818,000 tons of food fish in 1973 but, in 1997, were net importers of over four million tons of food fish. More than half of fishery exports originate in developing countries, while developed countries account for 84 percent of world fishery product imports. Fish products are vital exports for many diverse economies, but the top exporters of fish and fish products are Thailand, the United States, Norway, China, Denmark, and Taiwan (in that order). Imports into the three largest importing blocs, Japan, the EU, and the United States, accounted for over three-fourths of the world market in international fish trade in 1994. As will be addressed, although fish trade is important for many, it is hardly on equal levels or terms, and benefits are rarely bestowed on those to whom the resource matters most for food and livelihood security.

Fisheries: An Open Access Resource

Due in large part to the way it has traditionally been managed, marine fish has usually been seen and used as an open access resource, an in many ways endemic characteristic which has been the base of its dynamics of use, management, and exploitation. The term refers to a situation where no single user has to pay for access to a resource, nor does that user have exclusive rights to the resource or the right to prevent others from using it. Physical characteristics of the oceans, and the fish they contain, such as scale, interconnectedness, and mobility of fish stocks, contribute to this open access nature. As such, they have usually defied national borders and been logistically difficult or impossible to monitor and govern. Theoretically, this can lead to a "tragedy of the commons," where "the open-access nature of the fisheries allows more and more people to enter the fishing sector, which leads to the economic (and possibly biological) over-exploitation of the resources." Even after the resource begins to decline, producers will continue to increase their efforts until operating costs can no longer be covered. By this time, the stock of the resource may be seriously depleted. The common-access nature, which has long been a characteristic of fisheries, is the foundation on which many of the causes of current problems in the sector are based.

Root Global Causes of Fisheries Resource Depletion

The simultaneous, and highly linked, realities that too many fish are being caught and the global demand for fish is unsustainably high are largely seen to be the result of four root causes: population growth, growth in incomes, urbanization, and advancements in fishing technology.

Population growth has contributed to an increase in demand for fish by increasing the number of people who demand fish. Income growth has played a role because, as incomes increase, wealthier people's tastes change from preferring low-value to high-value fish; more and more low-value fish are then diverted "from poor people's plates into the stomachs of expensive carnivorous fish, which will in turn end up on the plates of wealthier people." A 2003 report by the World Fish Center found that population growth and income growth were "widespread structural phenomenon," especially in the developing world, which drove increased demand and consumption in the last three decades.

Urbanization has also impacted the structure of fish demand because, as the "rapidly expanding" group of poor people living in larger cities throughout the world increases, low-value fish is their only affordable source of animal protein. On the production side, developments in technology that have allowed fishermen to catch more per unit input and seek out fish in deeper waters have kept production costs down and allowed production levels to increase.

Roman Grynberg and Martin Tsamenyi, for instance, cited constantly improving technology as a fundamental cause of fish stock depletion. The interaction of these structural forces of increasing incomes, populations, and urbanization, as well as advancements in technology, have lead to the over-use of the resource, which is, as is currently being witnessed, limited in its capacity to respond to and accommodate them.

RESPONSES TO THE CHALLENGES OF FISHERIES USE: MANAGEMENT AND SUBSIDIES

The very reality that the root causes of the over-use of fish stocks are structural makes it nearly impossible to completely alleviate the pressures of rising demand, increasing abilities to pay for high-value species, and more effective ways to capture fish on an open-access resource. But people, organizations, and governments have responded to these realities in many ways, their responses sometimes improving the situation, but often exacerbating it. Two of the most prominent and important responses have been subsidies and management schemes. The first, subsidies, occur when trade measures directly influence the income or reduce the costs of production for industry. Management schemes, the second, encompass the ways in which fisheries are governed and monitored. In practice, the way this is done often highlights a lack of effective management. Subsidies and management, which can be seen as responses by individuals, groups, or governments to the root causes of fisheries depletion, are also the areas where the most opportunities for change in the way fisheries resources are used exist.

Effective management regimes are considered by many to be key in the sustainable use of fisheries resources. Theoretically, effective management encourages a sustainable use of resources through providing conservation regulations and incentives for responsible use of the resource. They are especially beneficial under the pressures of international integration; responses by producers to the economic incentives from trade liberalization, for example, can be better regulated when effective management schemes exist so that over-exploitation and destruction are less likely to occur. Unfortunately, the prevalence of such management has been scant in real-world fisheries. As the United Nations

Environment Programme (UNEP) reports, "very few fisheries management systems have demonstrated the ability to keep catches below levels that put pressure on stocks." As noted in a study by the World Conservation Union (IUCN), "national and international progress on trade expansion and liberalization outpaces progress on fisheries management and articulation of sustainable development strategies." Fisheries management, crucial to sustainable use of the world's fisheries, has been difficult to implement effectively and has often left fisheries open to influences from other sources such as state trade measures.

Subsidies, one of these trade measures, have played a proven role in aggravating the causes of fisheries depletion. Theoretically, subsidies can lead to overuse of a resource by altering the behaviour of producers. Revenue-enhancing or cost-reducing subsidies, such as those that will be examined in this paper, increase marginal profits at each level of effort and therefore theoretically lead to an increased overall effort. This may have the short-term effect of creating additional economic rent for producers, but either new entrants or increased effort by existing producers, also encouraged by the subsidy, will shift the level of effort so that rent is eventually dissipated. Assuming that the management program does not effectively impose a sustainable level of resource use, cost-reducing and revenue-enhancing subsidies will push the level of overcapacity and overall effort even further than would an open-access resource in the absence of subsidies. The main theoretical effects of cost-reducing, rent-increasing subsidies are increased capacity, a delay in exit by producers from the sector, and increased effort by producers. These effects can cause the overuse of a resource, which can have many other significant impacts such as threats to food, livelihood, and resource security.

FISHING SUBSIDIES

Types Used by National Governments

In the real world of governments, fishers, fisheries resources, and the incentives that link them all together, a subsidy can take many forms. Fishing subsidies currently account for large portions of many industrialized country budgets; a 1998 World Bank study estimated annual budgeted expenditures on global fisheries subsidies to be between $14 and $20 billion, and that subsidies account for 20 to 25 percent of the sector's revenues. One type of subsidy, domestic assistance programs, form a clear example of government endorsement of the fishing sector. These programs are defined as budgeted assistance to fishing production, and often enhance operations and capacity. Japan, one of the biggest subsidizers of fishing activities in the world, has used domestic assistance programs to recruit young fisherman, give aid to fish cooperatives and boat owners, support marketing, and encourage price stabilization.

A 1998 World Bank report estimated that Japan budgets $270 million a year for these types of programs, and a total of $750 when aid for fishing vessel insurance is included.

The European Union (EU), another major subsidizer, supports such structural programs as fleet renewal and modernization, processing and marketing, aquaculture, maintenance of port facilities, and genetic product promotion in their fisheries sector. In addition, it pays for market supports to the sector in the form of a minimum import price program and measures to support price floors. Costs of these programs to the EU in 1996 were estimated to be about $530 million. While estimates of exactly what types of domestic subsidy programs the Chinese government uses are hard to determine due to scant data, they have clearly aggressively promoted the expansion of their fishing sector in the last 15 years (and have become the world's leading producer) and are estimated to budget between $500 and $750 million annually for domestic subsidies. While other countries subsidize their domestic fishing industries, they have comparatively small budgets for such programs as so do so on a smaller scale. The shear magnitude of the domestic subsidies in industrialized countries suggests the possible impacts on the behaviour of fishers their presence may cause.

A less obvious, but yet potentially just as influential, type of subsidy occurs when a country pays the access fees for their fleets to fish in foreign waters. Unlike domestic subsidies, these payments are not made directly to fishers or even domestic programs, but to the foreign governments who have jurisdiction over the fish stocks. The EU is one of the major subsidizers of distant-water fishing through its numerous access agreements with developing countries, mostly in West Africa. For example, the EU signed a five-year access agreement with Mauritania in 1996 that lifted an EU embargo on fishery imports from Mauritania and specified EU payments of almost $350 million over five years to the Mauritanian government. In return, the Mauritanian government granted increased access for EU fishing vessels in their waters, authorization of higher EU total harvests, and the specific allowance of EU-directed fishing of highly valued squid and octopus. The monetary benefits to the "cash-strapped" nation were described as a "windfall;" in consequence, the number of eligible EU boats rose from 165 to 240 and allowable EU harvests from 76,050 to 183,392 tons. The EU has similar agreements with other West African nations, as well as others in East Africa, the Indian Ocean, and South America. Another distant-water fleet subsidizer, Japan, currently budgets about $200 million a year on distant water arrangements with foreign countries and foreign fisheries assistance programs (which often help secure fishing rights). These payments went mostly to developing countries in the Pacific Ocean and elsewhere. The subsidization of distant water fleets from industrialized nations can have serious effects on the coastal communities of the developing countries with whom these agreements are typically made.

Possible Impacts of Perverse Subsidization

The impacts of government subsidies of fishing are at least as complex and numerous as the types of subsidies themselves, but in the context of sustainable development those related to market access due to trade measures and resource, livelihood, and food security are the most relevant. The trade implications of the subsidization of the fishing industry arise when the lower costs producers see as a result of the subsidies repress world prices for fish products. Those producers, who are often from developing countries that cannot afford subsidies, are then excluded from the market because they cannot afford to produce at the world price.

For example, when Japan subsidizes its fleets to fish in the waters of Chile, Japanese fishers bring the fish back to Japan and sell them on their domestic market. Chilean fishers try to enter the Japanese market and sell these same Chilean fish to Japanese consumers, but are kept out of the market because their costs are higher and the prices on the Japanese market are too low. In effect, Japanese fishers push Chilean producers out of the market for the sale of their own fish. This example illustrates the way in which subsidies in the fishing sector can exclude those who are outside the national subsidy regime from participating in the market.

Due to numerous examples where the presence of subsidies seemed to increase the capacity in the sector and, in turn, deplete available resources, fishing subsidies have also been blamed for a tendency to encourage overcapacity and overfishing. The case of stock depletion in Canada's Northwest Atlantic fisheries in the 1950s and 1960s illustrates how subsidies can increase capacity and lead to depletion. Between 1954 and 1968, Canadian subsidies increased the capacity of the Northwest Atlantic offshore fishing fleet by more than 18 times, creating twice as much capacity as could capture fish if resources were to be used sustainably. In 1970, the Canadian government "acknowledged explicitly that its subsidies for vessel construction and modernization over the previous two decades had led to the rapid expansion of larger vessels, creating serious overcapitalisation." An Organization for Economic Cooperation and Development (OCED) study reported that subsidies also contributed to overcapacity in the fishing sector in many member states: New Zealand in the 1970s and early 1980s, Spain's Galician fisheries in the 1980s, Norway in the 1960s, the EU during the 1980s, and the United States (US) in the late 1970s and 1980s. These examples of the history of fishing fleet subsidization in industrialized countries and the resulting depletion of their stocks illustrates how overcapacity in a sector can encourage overuse of a fragile resource and lead to its exhaustion, posing threats to the surrounding ecosystem and human environment.

In addition to threatening the resources on which humans depend for their incomes, fishing subsidies may pose direct threats to crucial aspects of human and national development such as livelihood and food security. These threats are seen most strikingly in the interaction between developing nations and the industrialized countries with whom they sign access agreements.

Frequent realities of EU-West African access agreements, for example, include a disregard of African fishing regulations and international agreements by distant water fleets, an inability by African states to monitor activities or catch levels of EU fleets, a lack of direct linkage between access payments and money spent on fisheries-related activities, a lack of enforcement of management efforts by African governments due to a fear of losing compensations, and the challenge of depletion of valuable fisheries stocks. There are also often many adverse effects on African coastal communities such as threats to the livelihoods of artisanal fishers and food security of those who depend of local fish stocks for protein intake. While EU payments often constitute important sources of revenue for West African governments, the security of fishing resources, vital to so many aspects of these countries, are often threatened.

The case of one West African country that signed an access agreement with the EU, Senegal, illustrates how subsidies and trade agreements can lead to threats on food and livelihood security. The Senegalese government initially implemented subsidies as a way to encourage fish production to supply domestic consumers with a protein source; fish accounts for 75 percent of the population's protein needs. But the use of subsidies has since become incompatible with this goal since the end result has been a re-orientation towards encouraging the exportation of fisheries resources and agreements with the EU for access to Senegalese fisheries. The combination of domestic export subsidies, which lower the costs of production for those fishers who chose to export, and trade advantages granted under the Lomé Agreement of 1982, which authorized customs exemptions to products originating in African, Caribbean, and Pacific (ACP) countries and increased the competitiveness of Senegalese piscatorial products in the European market, have encouraged small-scale fishers in the country to both intensify their efforts in response to international demand as well as orient themselves towards the export market. These developments in subsidy policy, along with several fisheries agreements that have been signed with Japan and the EU, have threatened livelihood and food security in Senegal. The impact of growth in subsidy-driven exports has lead to a "breakdown in the supply of cheap protein to the population," since local producers began to switch their activities toward the capture of species of high market value in European economies rather than those affordable to local consumers. The increasing scarcity of fisheries resources has also affected the traditional, artisanal fishers, whose techniques, which require low technology and employ many people, are beneficial for local resource and employment security. Since fish have become scarcer, foreign fishing fleets have become a larger presence, and local fishers lack the equipment and fuel to travel farther out to sea where the remaining catches are located, they have been forced to make deals with European or Asian boats to remain employed. In return, locals allow the foreign boats access to areas originally reserved for their needs and use. This trend threatens local control of the resource as well as livelihood security. In a country where fish provides a vital source of sustenance for the local population, in terms of both food and employment, the way in which subsidies encourage a reorientation of resources to foreign markets is a matter of much concern for the sustainable development of the country.

A ROLE FOR THE WTO IN ACHIEVING SUSTAINABLE DEVELOPMENT OBJECTIVES?

The Environmentalist-Free-Trader Alliance

As the potential threats perverse subsidies pose to the world's fisheries and those who depend on them become increasingly evident, many in the international community are seeking ways to link reforms in subsidies to those clearly also needed in management and resource use. One option that has recently become popular among environmental and sustainable development organizations and several fish-producing nations is the pursuit of subsidy reform through the WTO. This endorsement is significant in several ways. First, the connection between the liberalization of trade through the WTO has not traditionally been a cause of environmental organizations. As Grynberg and Tsamenyi note, "What is peculiar is that environmentalists, who for so long focused on economic and population growth as the cause of resource depletion, should now enter into alliance with the free traders and focus on a symptom rather than what they have traditionally seen as the root cause of the problem." Environmentalists have also conventionally highlighted the numerous potential conflicts between the liberalization of international trade and the use of the world's resources. In the context of fishery reform, however, international conservation organizations such as the WWF and the IUCN have expressed that their organizations hold several aims parallel to those of the WTO. A related second shift this "alliance" illustrates is the progress of thought concerning the role of the WTO and trade liberalization. Instead of seeing measures such as subsidy reduction purely as ways in which to influence economic factors such as market access and prices, these same measures are now being seen as capable of influencing areas of sustainable development such as the environment and human and national development.

Since its creation in 1995, the WTO has included sustainable development as one of its objectives along with the liberalization of international trade, but the linkages between the two objectives have not always been clear. The WTO promotes a world trading system based on the principle of comparative advantage, suggesting that, "In a trading system in which all nations remove protectionist measures and other barriers to market access, the net benefits to all should increase." According to the WTO, subsidies are trade-distorting measures that impede the operation of the principle of comparative advantage. What the WTO, a forum for negotiation of trade measures, cannot assure is to whom the economic rents from this process go or how resources are used in the process of trade and production.

Background on the Place of Fisheries in the WTO

Fisheries, their trade, and the subsidization of their production have yet to find their places in the WTO. Due to conflicting interests during the Uruguay Round negotiations, the issue of fisheries subsidies was not resolved at the Organization's inception, and fisheries subsidies were included under the Agreement on Subsidies and Countervailing Measures (SCM). This agreement covers all goods except agriculture and attempts to place restrictions on the power of WTO members to provide subsidies to their industries. As UNEP reports, since 1995 a "significant change in the political context of the issue has taken place. The impetus for inclusion of fisheries subsidies on the international trade agenda has increased substantially, primarily because it has been treated not only as an issue of efficiency and equity in international trade but also as an issue of protecting natural resources from depletion." Momentum for getting fishing subsidies put on the trade agenda increased in the late 1990s, in particular at a special session of the General Council of the WTO where Iceland, with the support of nearly 20 other countries, proposed that members agree to eliminate those subsidies that contribute to overcapacity. The desire among many WTO members to eliminate capacity-inducing fishing subsidies set the stage for an increased role for fisheries in future WTO negotiations.

In anticipation of the fifth Ministerial Conference in Cancun, Mexico of September 2003, a group of states calling themselves "The Friends of the Fish" lead the attack on fishing subsidies and their apparent threats to world trade and sustainable development. The group, which formed prior to the Seattle Ministerial Conference of 1999, successfully campaigned for inclusion of a paragraph in the Conference's draft declaration but failed to see an agreement be reached as the meeting broke up without a final ministerial declaration. Australia, Chile, Ecuador, Iceland, New Zealand, Peru, the Philippines, and the US are the leaders of the "Friends." Japan and South Korea, which have large subsidy programs for their fishing fleets, have argued that fishing subsidies should not be treated in terms of having trade-distorting effects and question the linkages others claims the subsidies have with resource exploitation. The EU is between the two sides; traditionally having sided with Japan and South Korea, it has recently been less defensive of its subsidies. Hoping for better results in the Cancun negotiations of 2003, the "Friends" then submitted a proposal to the WTO in April of 2002 outlining the "potentially harmful trade, developmental and environmental effects of subsidies to the fisheries sector" and suggesting the need to address these effects. While negotiations at Cancun also failed to deliver a consensus on the issue, the group and their allies continue to push for reforms through the WTO that would eliminate what they consider perverse subsidies in the sector.

Input of International Organizations

The call for the reform of fishing subsidies has come not only from member states of the WTO, but also from international non-governmental organizations. Economic organizations that have become involved include the OECD, which undertook the first study to measure fisheries subsidies in OECD countries in the early 1990s, and the Asia-Pacific Economic Cooperation Forum (APEC), which collected data on fisheries sector support programs in member governments. The FAO has played a crucial role in putting fishing subsidies on the radar; by their account they "sounded the alarm about the negative effects that subsidies were having on capture fisheries" in 1992, which "led governments of major fishing nations and various Intergovernmental Organizations to look into the issue." These organizations stress the link between fishing subsidies and their ability to cause perverse economic incentives and overcapacity.

Organizations with conservation and sustainable development agendas have also added their voices to the criticisms of fishing subsidies, and many have iterated a possible role for the WTO in the attack. The WWF, for example, has published much literature emphasizes the link between government subsidy payments and overfishing and, in one essay, suggests that "The World Trade Organization (WTO) today has an unprecedented opportunity to help improve the environmental and economic health of the world's oceans by disciplining harmful fishing subsidies." The IUCN has also described a role for subsidy reform and the WTO in the achievement of fishery use within sustainable development aims. One publication examines possible "synergies" between "international trade, sustainable fisheries, the environment, and sustainable development." By focusing on the causal factor of subsidies in overfishing and resource depletion, these organizations have created a bridge between a trade measure and the state of natural resources, thereby creating a role for an economic organization such as the WTO in the pursuit of sustainable development aims.

The Sustainable Development Concept

The context into which governments and organizations are currently trying to put subsidy reform, that of sustainable development, pulls together concerns that are a combination of economic, developmental, social, and environmental. The concept of sustainable development became prominent at the 1992 United Nations Conference on Environment and Development, which defined the idea as "meet[ing] the needs of the present without compromising the ability of future generations to meet their own needs." With its incorporation of a commitment to poverty eradication, reductions in disparities of standards of living, and environmental protection as an integral part of the development process, sustainable development "challenges the system to reconcile trade objectives with broader environmental, social, and cultural imperatives, and also raises concerns as to the issue of what constituted effective and legitimate development, rather than measuring development in narrow terms of economic growth." Sustainable development in terms of fisheries includes meeting the economic, social, and cultural needs of coastal communities, addressing the concerns of developing countries to maintain revenues from trade that are necessary for development, monitoring the health of fish stocks, and surveying the surrounding situation of biodiversity. This perspective encompasses much more than either the strictly economic side-benefits to increased fish production and trade, or the conservation side-sizes of fish populations.

Reform of fishing subsidies, an issue that seems to have become a nexus of economic, environmental, developmental, and social interests, is defined by many as a potential 'win-win-win' situation for all concerned. According to the WFF, "the WTO and its members have been promising for years to negotiate new rules that deliver true 'win-win-win' outcomes for trade, the environment, and sustainable development," and the fisheries subsidy issue presents the "first real" opportunity to move toward this objective. In suggesting that such a situation exits, 'win-win-win' outcome proponents conjecture that the same measure, here the elimination of perverse fishing subsidies, could deliver favourable benefits to trade by increasing market access for producers, the environment by reducing the capacity of global fishing fleets, and sustainable development by assuring resource and livelihood security to those in developing countries.

The Argument for the Use of Subsidy Reform in the WTO to Achieve Sustainable Development Objectives

The argument for proposed changes brought through the WTO, that many view as having the potential to deliver this 'win-win-win' outcome, are based on three main justifications of subsidy elimination: trade, sustainable development, and the environment. The trade foundation of the argument deals with the way in which cost-decreasing, rent-increasing subsidies may deflate prices on the international trading market because of lower costs for producers due to government support that absorbs a portion of costs. The submission made by the "Friends of the Fish" to the WTO stated that subsidies in the sector caused distortions to trade through their "effects on prices and production levels," and that the transfers "significantly distort trade in fisheries products by lowering costs of subsidized production, increasing the volume of subsidized production and thus distorting competitive relationships in markets for fisheries products." Concerns for the impact of such distortions are also voiced by organizations, mainly in the context of how they might adversely affect the economies of developing countries. An IUCN report suggests that, for many developing nations, "trade liberalization offers possibilities for addressing declining terms of trade," and that efforts to reduce subsidies could have the positive impact of enhancing " export opportunities for countries currently shut out of export markets due to subsidized competition." These factors are important for developing countries in particular because of questions of equity in trade; as Charlie Arden Clark of the UNEP expressed, those best able to subsidize themselves (wealthy developed countries) will best be able to obtain the resource and trade it on the world market. The importance of fishery product exports to the economies of many developing countries is also an important part of the argument; the FAO highlights how "exports of fishery products are vital to the Least Developed Food Deficit Countries, in terms of export revenue, employment and income generation." The potential benefits to elimination of perverse subsidies in the fisheries sector such as increased market access for non-subsidized competitors comprise part of the argument in how subsidy reform might form an important part of sustainable development.

The potential positive impacts of addressing fishing subsidies in the context of an international trading system could extend beyond those that are purely economic, though, and contribute to several important aspects of human and national development. Due to the way in which they can threaten food and livelihood security, the effects of access agreements and the foreign fleets they bring into the waters of developing countries are being targeted in particular as a way to move toward development aims. An IUCN report notes that one positive impact might be to "Provide greater opportunities for domestic fisheries industries in developing countries by reducing subsidies to foreign distant water fleets." If access agreements were eliminated, there may be less competitive pressure from foreign industrial fleets, which "effectively compete for fish with local artisanal and industrial fisherman, thereby threatening their livelihood security, and undermine the economic development of poorer West African nations." The presence of foreign fleets and the governmental revenue they bring are also of little development assistance to local fishing communities. As Aimee Gonzales of the WWF explained, the fees that developing country governments receive as access payments are not used to access capacity or monitor stocks, measures that might enhance local capacity to develop a sustainable fishing sector. She also described how there is little "trickle-down" effect from the presence of the industrial fleets in the coastal communities; since the big, foreign boats have the ability to process their catches on-board rather than use local processing plants, they contribute little to the local economy. The elimination of access agreements may, in some cases, also help increase food security by increasing the share of fish that go to foreign fleets for export. Developing countries hypothetically stand to gain much from a decreased presence of foreign fleets in their waters.

These potential gains for development in eliminating foreign access to fisheries of developing countries have inspired proposals from several organizations for reform. An IUCN report, for example, challenges the agreements to be considered subsidies under the WTO SCM Agreement, as well as proposes a new international agreement that "includes provisions prohibiting the payment by governments of any part of the costs of access to fishery resources." The WWF also recommends that the WTO expand its definition of "subsidy" to include the access agreements, and to address the issue of sustainability by "considering whether subsidies contribute to excess capacity, over-fishing, or unsustainable fishing practices." Under such propositions, a subsidy may be prohibited under WTO rules, even if it does not cause trade-distortions, as long as it is seen as being counter-active to sustainable development goals.

The environmental impacts of cost-reducing, rent-increasing subsidies are looked at mostly in terms of how the subsidies contribute to fishing fleet capacity. The link between subsidies and overcapacity is crucial in understanding both the role subsidies have played in causing the current state of resource depletion and what role they might play in the future in the pursuit of more sustainable fishing behaviour. According to an UNEP study on the effects of subsidies in the fishing sector, "The key issue for analysis and assessment in regard to the linkage between subsidies and overfishing is whether and to what extent fisheries subsidies contribute to fleet overcapacity." Fishing fleet overcapacity has been identified as a major threat to fishery resources, and global fishing fleets are currently estimated to be up to 250 percent greater than needed to catch what the world's oceans can sustainably produce. Carl Gustaf Lundin, Head of the Marine Programme at the IUCN, emphasized the need to reduce capacity in order to achieve fisheries resource security. Because of the threats they pose to the environment, governments and organizations suggest that capacity-inducing subsidies be prohibited by the WTO. An IUCN report proposes that various modifications be made to the WTO SCM Agreement to "ensure that all capacity inducing subsidies are covered by the agreement." The WWF and the "Friends of the Fish" also suggest that the WTO rules address the issue of subsidies that contribute to excess capacity and contribute to fish stock depletion. These groups hope that, if subsidies that increase fishing fleet capacity are targeted and eliminated, fish stocks may be able to recover in the face of decreased fishing intensity.

While most of the argument for the link between the elimination of perverse subsidies and certain sustainable development goals has been based on these three objectives, some feel that a role for the WTO may in itself add to progress. They believe that the WTO can potentially increase political action in the area of fisheries and improve other important aspects that need reform, such as fisheries management. Aimee Gonzales at the WWF explained that, since subsidies are where the politics are, action through the WTO might spur "parallel" action among other bodies to create good management regimes and, thus, sustainability. Hugo Cameron of the International Center for Trade and Sustainable Development (ICTSD) agreed with this idea, noting that the fact that the subsidies issue is being brought up at the WTO will give fisheries "air time." He also felt that recent improvements in the EU's Common Fisheries Policy were due in part to pressure from governments and environmental groups for reform inline with WTO rules. The "Friends of the Fish" submission to the WTO describes a role for the WTO in the context of a situation where action is clearly needed:

While many factors contribute to depletion of the world's fish stocks, harmful subsidies are one important factor within the WTO's competence to address....As the forum in which global trade rules are negotiated and enforced, the WTO has a unique and important role to play as part of a multi-layered global response. The WTO's mandate, experience, expertise, and pre-existing framework of rules and dispute settlement mechanisms are critical to dealing effectively with the trade-related aspects of this problem.

An integral justification of the proposed role for subsidy reform through the WTO has become the benefits the WTO can uniquely contribute to the situation due to its international political importance and prominence.

Evaluation: The Implications of WTO Subsidy Reform as a Method for Achieving Sustainable Development

Given the complex and far-reaching impacts organizations and states conjecture for fishing subsidy elimination through the WTO, it will be difficult for actual reforms to live up to their model. Certain benefits seem assured. On a net basis, trade liberalization can increase the economic gains of a developing country by increasing their market access. And, in a vacuum, the elimination of subsidies can decrease productive capacity and leave more resources available for the use of those who were formerly excluded. Sustainable development, though, is a much taller order. It means assuring that certain aims can be met in the context of forces such as trade and resource use. Whether or not WTO action on fishing subsidies can achieve these aims is the question. The ability of subsidy elimination to decrease overfishing by subsidized fleets or create market access where there perhaps was none before is not under debate-what are are the implications and possible benefits of this action.

Regarding the environmental sphere of the argument-the use of subsidy reform for conservation aims-the security of fisheries resources has many factors beyond that of the subsidization of fleets. While the theoretical reasons that support the role for cost-reducing, rent-increasing subsidies as a potential cause of increases in capacity are strong, and many empirical cases have illustrated instances when increases in such subsidies played a pivotal role in overcapacity, the extent to which they have done this is still under debate. The importance of management in the relationship between subsidization and overfishing has been found to be incredibly significant since effective management may alter the impact of subsidies on resource use. If fisheries are well managed and impose regulations and/or quotas, additions to capacity may not result in additional effort. The OECD Committee for Fisheries reports that the effects of subsidies on fisheries resources are theoretically different depending on the type of management regime present. As could be expected, the effects of subsidies on open-access-type management regimes include overexploitation and increased capital and labour being attracted to the industry. In contrast, under an effective management scheme, the same subsidies had no effect on resource sustainability. By conjecture, if effective management schemes were also present in developing countries, the degradation of their resources in the face of highly subsidized fleets may not seem so imminent. Theoretically perverse subsidies clearly create overcapacity and resource degradation in many cases, but do not stand alone as the causal factor of current fisheries depletion.

The validation for subsidy elimination based on trade illustrates the complexity of achieving a true 'win-win-win' situation, especially when the interests of one of the main supporters of this view, the "Friends of the Fish," are examined. The group, described by Hugo Cameron at the ICTSD as a group of relatively competitive fishing countries, is lead by the US and New Zealand, fishing nations that subsidize their fleets relatively little. Support by these two countries in particular seems to be rooted in their economic concerns over the competitiveness of their fishing fleets. The US government apparently blames the lack of profitability of its fleets on the low world fish market prices, which it sees as being held down by the subsidies in distant water fishing nations. Eliminating subsidies in the EU and Japan, for example, would theoretically increase prices and improve the US's market share. According to Grynberg and Tsamenti, environmentalists who support trade liberalization as a means to sustainable development and environmental security are in effect supporting US trade interests. To the "Friends of the Fish," eliminating subsidies in the fishing sector is a way to increase the competitiveness and productivity of their fleets.

What the motives of the US and the "Friends" highlight is that there is a significant endorsement of free market principles in the attack on fishing subsidies, principles that, if implemented, might unleash threats on areas crucial to the pursuit of sustainable development and resource security. An IUCN report suggests the possibility that "the free functioning of comparative advantage in trade relations may not per se encourage the best long-term allocation of resources in the fisheries sector." This possibility becomes especially acute when the lack of effective management of developing country fisheries is recognized. While, in many industrialized countries stocks are monitored and assessed and few fisheries remain purely "open access," developing countries lack the capacity and funds to effectively manage their fisheries. The WFF reports, for instance, that "Traditional management methods like surveillance patrols, ecological monitoring, or measures to address pollution from both land and sea have met with only limited success because they are simply too costly for the region's cash-strapped institutions." This asymmetry in management between the North and the South may actually encourage fishing in developing country waters, where management is more lax. This idea is supported by economic theory, which warns of a comparative advantage that can seem to be present for the production of common property resources in the South due to their less well-defined management or property rights schemes. As described by the economist Graciela Chichilnisky, "lack of property rights alone can create trade" between two countries that are otherwise identical since the ill-defined property rights in one country alone will cause the overuse of a resource in that country. Fisheries biologist David Boyer explained that increased market access for developing countries may indeed increase pressure on fisheries as fishers in these countries scale-up their fleets in response to international demand for high-valued fish. In his opinion, market liberalization may well "shift pressure to greater fishing for trade from countries that have little environmental protection and the inability to enforce international treaties or their own national laws." He also notes that, without "deliberate domestic policies to facilitate restructuring" in the face of trade liberalization, "liberalization may, at least in the short and medium term, actually work against growth, employment, poverty alleviation, and other components of sustainable development." Liberalization of fish trade, one of the impacts of subsidy reform, may only continue the current pattern of resource overuse that threatens the development and resource security of developing coastal nations, or, at worst, increase this overuse.

Better management of fisheries in developing countries increases the chances of "winning" on several development fronts since it provides some assurance of how the resource would be used after the elimination of subsidies in addition to "less," or "more in line with market principles," as the subsidy elimination argument suggests. As described, free trade of fisheries resources may put pressure on the resource and do little to assure livelihood security for local communities. Effective management schemes could enforce quotas and regulations, assuring livelihood security for future generations. Current livelihood security in many developing fishing nations further depends on the lessening of competitive pressure from distant-water fishing fleets from developed nations. Subsidy elimination would very likely lessen the number of EU or Japanese boats in these nations' waters, but this change says little about which boats, such as those of Iceland or Argentina, known to be very efficient, might enter next. Even when subsidies are removed, high global demand for fish will still exist, as will the reality that a large percentage of the world's fish are located in the waters of Western Africa and Western South America. In my opinion, management is one of the only likely mechanisms that can relieve such market pressures on the resource. Food security may be improved little by removing subsidies if, as the trade scenario described suggests, producers still have large incentives to focus their efforts on high-value fish for export to developed countries. If management were effective, though, catches by industrial or foreign fleets could be limited and some resources could be reserved for local population use, a method that has been proposed in Mauritania. Management of fisheries in developing countries is not only a key aspect of resource sustainability, but also a necessary guard against potential threats of subsidy elimination.

Organizations and the "Friends of the Fish" argue that certain characteristics of the WTO, such as its prominence in political and economic arenas, may help sustainable development aims by leading to a chain reaction of reforms in other areas. This argument's merits aside, the international scale of the WTO, whose rules "are intended to help ensure that trade flows as smoothly, predictably, and freely as possible" at an international scale, poses potential threats to interactions at other levels. For example, as the push for action on fisheries issues at the WTO increases, those engaged in fisheries agreements at the regional level are raising concerns about potential conflicts between regional agreements and international WTO rulings. Regional Fisheries Organizations (RFOs) adopt their own trade measures between states in a region in order to achieve purposes such as compliance with conservation and management regimes. RFOs can ensure cooperation through requirements for documentation, prohibiting landings from non-compliance vessels, or implementing trade measures. Given their aims for effective management, RFOs present an opportunity for reform in an arena crucial to fisheries resource conservation and largely left out of the impacts of subsidy reform. But, as Richard Tarasofsky forewarns, RFOs, which "seek to discriminate in favour of goods produced in accordance with their conservation and management aims, are potentially in conflict with these [WTO] provisions," which mandate equal trade treatment of like products. A winning situation at the WTO may be a loss at the regional level if WTO rules that assure the smooth international trade of equally treated products compromise the ability of regional organizations to institute effective management schemes and assure resource conservation.

The threats the international focus of the WTO pose for RFOs are an example of what action on fisheries subsidies at the WTO could mean for developing country national governance. Governance of fish stocks currently rests at the national level-since the UN Law of the Sea Convention of 1982 states have had sovereignty over stocks that lie within 200 nautical miles of their shores. As the subsidy reform issue highlights, management is a key aspect of development and resource security in developing coastal nations. National governance is thus an unavoidable part of a solution and holds many opportunities for conservation of fisheries resources by intervening in the domestic situation and creating management solutions. The government of Senegal, for example, announced in September 2003 the creation of four marine protected areas that are of vital importance to local fishers and the domestic economy. While this action was hardly a subsidy, at the level of the WTO some states may argue that any intervention by a government in the fish sector destroys the market. Hugo Cameron of the ICTSD gave a current example of this argument when he explained how, although many states are trying to use subsidies to reduce capacity by using subsidy funds to buy-back vessels or employ people to repair ports, states such as those in the "Friends of the Fish" may argue that this still provides people an incentive to remain in the industry and distorts the market. As Carl Gustav Lundin warned, it can be a "slippery slope" when international bodies limit the interventions of national governments. In order to achieve sustainable development objectives, WTO rulings on fisheries must not interfere with developing nations abilities to use government support measures in their domestic economies and resource management sectors.

Although governance of fish stocks rests at the national level, sustainable development objectives require that livelihood and food security be met at the local level, and that future solutions do not stop at questions of national access or resource use. Conservation, fair economic activity, and development are meant to be non-national goals, even though, as the fisheries subsidies issue highlights, they have often been dealt with in terms of national policy. Ending the analysis at the national level, though, is a fault. Current threats in the fisheries sector at the local level come from the way in which trade liberalization through the WTO may not assure that the distribution of benefits from export revenues correlate to any measure of development objectives. Increases in national export revenue do not necessarily represent changes in people's access to resources for food or livelihood. Instead, they may simply reinforce the existing economic structure within the community. This is especially so in this case due to the lack of effective management of fisheries resources, which, as described, may only add to resource depletion under trade liberalization. Given the way the WTO does not monitor such local dimensions, it is incapable of assuring any outcome outside that which relates to the state as a single economic entity. A real resolution to the crisis would take the inner view of the national situation-most importantly here the level of resource management-into account and tailor policy solutions to this.

Conclusion

In order to achieve a true 'win-win-win' solution, subsidy reform must deliver not only benefits on the international level; global levelling of trade advantages in the economic sphere, net conservation of resources in the environmental sphere, and a targeting of resources toward development aims, but gains at all levels. This is perhaps the most challenging aspect of the role proposed for the WTO, a fundamentally international organization. In order to be successful, it needs to prove that it can improve the lives of people down to the local level. The concepts of sustainable development are an entry point into such progress, but, as this paper describes, there are serious limitations to relying on an international economic institution to deliver such goals. The threats which have been outlined; the lack of assurance that gains at the national level through increased export revenues and trade liberalization will deliver resource or livelihood security, the threat to national autonomy due to WTO intervention in the responses of state governments to the fisheries crisis, and the dire need for effective management in the fisheries of developing countries, represent needs on which WTO action as it is presently seen cannot fully deliver. Only when proposed reforms can assure that objectives are met in the three pillars of sustainable development-environment, trade, and development-as well as at all levels of organization and governance can any outcome be considered to be winning.

FISH BIODIVERSITY AND PROFIT MAXIMISATION: DOES A TRADEOFF EXIST?

Introduction

The environmentally sustainable use of fish resources is central to fisheries management, given the long-term importance of this sector in terms of nutrition and employment. But today's major concern relates to the unsustainable levels of exploiting fishes with such practices that lead to the depletion of fish stocks, disruption of ecological equilibrium and reduction in diversity. A recent re-analysis of FAO catches statistics (Pauly et al., 1998) documents progressive 'fishing down' of food chains as fishing effort responds to depletion of original target stocks where one species is exploited more than another their relative abundance changes, as in the North Atlantic (Sherman, 1990). This is referred to in the literature as 'fishing down the food web'. Virtually all commercially valuable marine populations are now overexploited. Overexploitation diminishes species population and reduces economic return. As the most valuable species are overfished, they are quickly replaced by catches of less desirable ones. It is seen that a large share of today's global catch consists of previously unused, less valuable species. This type of phenomenon has been identified as 'fishing down the value chain' (Kasulo and Perrings, 2001) in the literature. 'Fishing down the food web' implies that the value of fished stock decreases, but this may not be captured until they make it to the market. Whatever be the fishing sequence, demand or supply driven, it will have an impact on fish biodiversity with a change in composition and relative abundance of harvested species. So a desire to increase profits may hamper economic biodiversity conservation and thereby affect the value of the fishery.

Case Study

Our study uses fish species as indicators of trends in coastal biodiversity. This study looks into the loss of fish biodiversity and its effect on profitability of a fishery. The theoretical aspect is modelled in an extension of the aggregated Gordon-Schaefer standard fisheries model. This study focuses on the Digha fishery in West Bengal on the eastern marine coast of India on the Bay of Bengal as a micro case study in an attempt to analyze the relationship between biodiversity and profitability present there, if any.

Our study area is a part of the coastal area of West Bengal. The total coastal area of West Bengal stretches between the mouths of rivers Herobhanga or Harinbhanga on the Indo-Bangladesh border in the east and Subanarekha in the west, the total length of which is about 220 kms. Quite naturally, fishing activities in this zone provide economic sustenance and a source of livelihood to a cross-section of people who, in turn support the flourishing trade in this lower Ganga deltaic region. The coastline of West Bengal spreads over two maritime districts-24 Parganas (South) and East Midnapore. There are 13 marine fish landing centers in 24 Parganas and 27 in East Midnapore.

The Contai coastal belt under the district of East Midnapore is considered to be highly potential in respect of marine fisheries activities. The coastline stretches from Digha under Ramnagar-I Block to Talpati Ghat under Khesuri-II Block and is about 60 km. in length. Digha is situated close to the Gangetic mouth on the east of India at lat. 210 36'N. and long. 870 30'E. It is located in the West Midnapore district of the State of West Bengal of Eastern India and lies in the southern most part of the state on the bank of Bay of Bengal. With the introduction of diesel using powerboats, deep-sea fishing and mechanization in fishing is taking an upturn. It has been observed in Digha coastal areas that total marine fish landing mainly consists of sardine, hilsa, coila, pomphret, croakers, Bombay duck, catfish, ribbon fish, shark, shankar, prawn etc.

Thus total 37 varieties of fish are found here. These varieties have been divided into five groups considering their importance from the viewpoint of their demand and price. Among them contribution of hilsa in total catch per trip was found to be maximum in Digha. It was followed by two types of pomfret: Chinese and Silver pomfret being one variety and the other being Black pomfret. So we see from collected data that mainly four varieties of marine fish dominate the Digha fishing industry in terms of both prices and quantity. They are Hilsa, Chinese and Silver Pomfret, Black Pomfret and Prawn. More than 50% of the total value of catch was contributed by these four species. Individual contributions of other 33 species in terms of value are not very significant. Also these other 33 varieties of fish such as (in local vocabulary) sardine, mackreal, chela, para and American bhetki have a very low price range in the market. These have been grouped under the heading 'others'.

The issue of investigating any relationship between biodiversity and profitability has largely been neglected in the literature for this particular location. The sustainability issue has been dealt in some empirical works pertaining to Digha fishery and they are essentially biological and static in nature (Central Marine Fisheries Research Institute, 1984; Guha, Neogi, Das and Chakraborty, 1994, Das, Neogi and Chakraborty, 1996; Das, Neogi and Chakraborty, 2000; Dhar, 2004, Jana, 2004). The Digha Fishermen and Trader's Association regulates fishing activities in that area and acts as a profit-maximising unit within a larger competitive fish market. But a void exists in terms of bioeconomic modeling of the fishery in a dynamic profit-maximisation set-up in the literature. Also the question of biodiversity needs to be addressed in the light of various reports of the marine fisheries of West Bengal. So this paper focuses on the dynamics of the profit-maximising regime and estimates the dynamic maximum economic yield and net present-value of fishery profit that is maximized here. Sensitivity analysis with respect to small perturbation in discount rate examines its impact on profit under different biodiversity scenarios. In the end we examine whether economic biodiversity conservation are in conflict with profitability of the fishery.

Importance of Biodiversity

The importance of biodiversity had long been recognized by ecologists and environmentalists. But the main focus came into being with the initiation of the Convention on Biological Diversity in 1992 (UNCED, 1992). The emphasis was on conserving biodiversity to achieve the sustainable use of its components and therefore to secure the fair and equitable sharing of the resources which that biodiversity represents. Our study relates to that area of the Convention that emphasizes on "... marine and other aquatic ecosystems, and the ecological complexes of which they are part.........." and the value that they represent.

Conservation of biodiversity is important in environmental management programs (Turner and Gardner, 1991; Reid and Miller, 1989). Measuring biodiversity is one of the central issues in ecology because of its importance in devising conservation strategies. However, incorporation of biodiversity in a fisheries model is a difficult task. This difficulty arises from the lack of a single practical operational definition and measurement of biodiversity. A simple measure in the literature of ecology used for quantifying fish diversity is to count the number of species in the habitat or community. This measure is considered to be too simplistic and so other indices measuring biodiversity based both on number and abundance of species have been identified. The idea of species diversity has two distinct concepts: species richness and abundance. Species richness is the number of species in a community, while abundance explains the relative proportion of each species in the community. The Shannon index is a very well known example of the richness index group. An important example of dominance index is the Simpson's index.

The main difference between the above two type of indices is that species richness indices are weighted more towards uncommon species while dominance indices are weighted more towards the abundant species. The question now is one of selecting the most appropriate method of measuring diversity. The choice of the index should depend on the component of diversity being measured. If the main concern is on rare species, then richness-based species should be used. But if the major interest lies in the abundance of the commonest species, then dominance indices are preferred. In our case of Digha fishery, data on catch per species will be used to compute the level of biodiversity. Here dominance indices is the most appropriate measure as catch data change shows changes in the abundance of dominant species. In order to apply the Simpson index to the case of fisheries, the unit of measurement must be changed from individuals in a sample population to biomass in the total catch (Goda and Matsuoka, 1986). So the Simpson index (henceforth called the unweighted or ecological Simpson's index) can be expressed as:

$$D_t = \sum_{i=1}^{s} (Y_{it}/Y_t)^2$$

where Yit is the catch of the ith species harvested in period t, Yt is the total catch in period t and s is the number of species harvested in period t. The use of this estimate enables us to identify an empirical relation between the diversity index and total catch. These diversity indices are basically ecological in nature as they emphasize on ecological differences among species that comprise a community or habitat.

Heywood (1995) reports about the growing perception among both ecologists and economists that the importance of biodiversity lies first and foremost in its role in the production of goods and services that are useful for human welfare i.e., in its social-economic importance. An individual harvested species is valued for specific properties that make it useful in either production or consumption. The biodiversity that supports such species derives its value from this. Any measure of diversity should accordingly reflect this (Perrings, 2000, Heywood, 1995). Ecological and economic values are not necessarily the same. It does not follow that if biodiversity is important to the functioning of some ecological system then it will automatically be valuable for society. Nor does it follow that a species that is rare will be economically scarce and hence valuable. So, there is no necessary correspondence between ecological biodiversity measures and economic value of biodiversity. Systems with a high biodiversity value by any of the standard indices may or may not have high economic value. So, to reflect the socio-economic value of diversity, the ecological diversity measures need to be modified (Kasulo and Perrings, 2001). In fisheries, as most of the species caught are marketed, an economic value of a species can be approximated by using market prices (Hanemann, 1988). The rationale for social-economic valuation of biodiversity lies in the fact that the signals generated by the market system i.e., prices lead to excessive rates of biodiversity loss (Heywood, 1995). Market prices influence the exploitation of a fishery and hence the level and direction of effort. Targeting of effort towards particular species leads to elimination of highly valued species, and to a reduction in biodiversity and productivity (Barbier et al., 1994). To capture the economic value of species, the Simpson's biodiversity index is modified such that it uses market values of species caught rather than the total amount of species caught. Now, the actual amount of the species will be weighted by price. Therefore, the Simpson's index becomes (henceforth called the weighted or economic Simpson's index):

$$B = \sum_{i=1}^{s} (P_i Y_i / TR)^2$$

where B is the economic biodiversity index, Pi is the per unit price of species i, and TR is the total revenue. When all the species have the same market value, the solution for economic biodiversity index is the same as the ecological biodiversity index. When the community is dominated by species of high market value, economic biodiversity index will be greater than an ecological biodiversity index of the same community and vice versa.

Ecologically, it is acknowledged that when the major concern is about the uniqueness of species, then each different type of species should have equal inherent value. However, when economic considerations are taken into account, then different species are assumed to have different values. Biodiversity may take on different values depending on people's perception of the economic or ecological importance of species diversity to human welfare (Barbier et al., 1994). In our study following the work by Kasulo and Perrings (2001) biodiversity is weighted by its economic significance. This is important because it reflects the value of biodiversity to humans. Thus, more weight is given to species with higher market values.

Data Analysis

Data for this study has has been collected from the Digha Fishermen and Fish Traders' Association in Digha covering the period 1993-94 to 2002-03. Catch is measured as kg.of fish landed and effort is represented by fishing months. Digha fishery, in recent years, has seen a shift in fish species harvested towards catches of fish species of very low local value (ranging between Indian Rs. 4-Rs.35 per kg.) consisting of sardine, chela and kaante which we have clubbed under the heading 'others'. This transition in fish catch from high valued to low valued species points to the role of the market and the effects of economic forces in loss of biodiversity. The decline in the dominance of hilsa in total catch reflects not only a decline in the trophic level of fishes but can also be associated with its economic value. The total landing of hilsa has declined to 1% in 2002-2003 whereas its contribution was 34% in 1993-1994. In contrast, the total landing of the species such as sardine, chhela and kaante has increased from 43% in 1993-94 to 75% in 2002-2003. It is interesting to note in this context that hilsa, which is a very popular traditional fish, has a high average price of Rs.73.18 per kg. while chhela, sardine and kaante are valued at very low average prices of Rs. 9.77 per kg.

Fig. 1: Comparison between ecological and economic biodiversity indices in Digha fishery

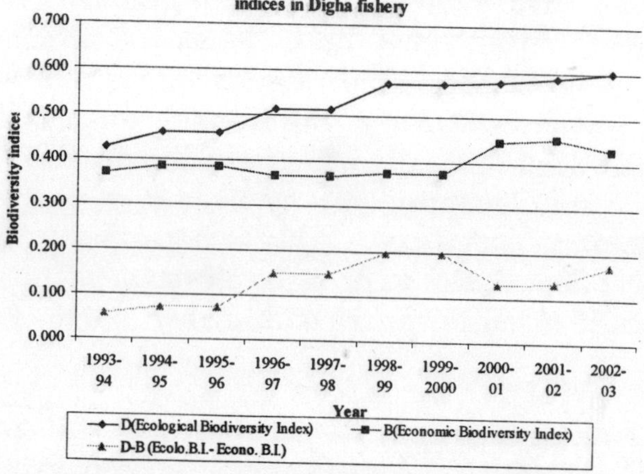

The market price in this case reflects people's preference and is one of the reasons for the over-exploitation of hilsa. So, not only is there a shift in relative dominance of fish species in total catch, it also reflects an exploitation from valuable to less valuable species. A comparative analysis of the unweighted and weighted Simpson indices is carried out by using the data on catch per species for the fishery of Digha. The Simpson economic biodiversity is constructed by weighting the simple ecological Simpson's index by average prices so as to capture fluctuations in value. It will capture any shift that may occur in fish value resulting from the over-exploitation of high-valued species. Data shows that a comparison between the Simpson unweighted and weighted indices, we find the value of the weighted index is lower than that of the unweighted index. It is because of the differences in the value of the species caught that the differences in the two indices occur. The lower values of the weighted indices in comparison with the unweighted index reflects that on average catches are dominated by less valuable species. If catches had been dominated by valuable species, price weighting would increase their dominance even further and the weighted indices would have higher values than unweighted indices (Kasulo and Perrings, 2001).

This is seen in figure 1 where it suggests a decline in economic biodiversity associated with a shift in fish catch from high-valued to low-valued species. Curve D reflects the ecological biodiversity index over the years and Curve B the economic biodiversity index over the period 1993-2003. The third curve D-B plots the difference between the ecological and economic biodiversity indices over the same period. The curve D-B takes on the shape of a somewhat inverted U-shape with the difference peaking in the year 1998-1999. In the early years of our study like 1993-1994 we find that the D-B curve is more or less flat shape at a low level, implying that the difference between the D and the B curves is fixed at a low level.

The implication is simple. At the initial stages of the liberalization regime we find that fishing was not much mechanized in Digha due to which fishermen there used to catch fish by using country boats or non-mechanised fishing boats. Naturally the cost of fishing at that time was not as high as we find in present situation. Due to lack of mechanized trawlers the fishermen at that time were unable to enter deep sea as a result of which they used to sell more or less similar type of fish species in the market. This situation is specific for the local markets near Digha. Apart from this, fish marketing in Digha was not as much developed as we find in present day situation. All these factors led to not much fish price per species variation along with its low price. Hence it can be argued that during the initial years of liberalization the difference between ecological and economic biodiversity indices was not much prominent. The gradual fall-off of the difference in the next few years show the loss of economic biodiversity that the fishery is facing with the more expensive fishes being fished out.

The socio-economic impact of loss of fish biodiversity when associated with a shift from high valued to low valued species are generally negative. Most of the fishermen operating in Digha coastal area are socio-economically backward with average literacy rate, given by census 1991, at only 30.68%. According to Digha Development Authority, there are 600 countryboat owners, 100 motorboat owners and 400 trawler owners operating in the fishery. The average family size in Digha is 4.74 persons/family (census 1991). The marine fishermen in Digha are mostly local people with negligible proportion of migrants, about 4%. With respect to land holding, average land per fisherman is very small. The primary reason behind this is that they find fishing business more profitable than cultivation. They work full time in fishing and do not have practically any other source of income. The shift in fish catch from high-valued to low-valued species means lower profits for the fishing vessel owners and hence crew labourers hired on per trip basis run the risk of losing their jobs. The trawler owners invest large amount of capital on boat and net while the countryboat owners, having non-mechanised boats, invests the least. Since these fishermen have poor economic background, they mostly have to borrow mainly from cooperative banks and private sources such as moneylenders called the Aratdars and Mahajans. A loss in fish diversity (reflected through fall in value of catch) will reduce profits and will mean an additional pressure in repayment and a subsequent debt trap for them. On the other hand, the crew labourers running the risk of unemployment and having no alternative source of income becomes worse off. In a survey conducted on Digha marine fishermen, Guha and Neogy (1996) has found that lower income group of fishermen spend approximately 60% of their monthly expenditure on food while higher income group spends 52% on an average on food items. So it is the crew labourer fishermen who are likely to be hardest hit by the decline in fish diversity.

COOPERATION IN A CONTEXT OF CRISIS: PUBLIC-PRIVATE MANAGEMENT OF MARINE FISHERIES IN SOUTH ASIA

Marine fisheries off the Kerala coast are on a turning point. Resource degradation, livelihood vulnerability and the growing threats of liberalization demand for an appropriate management regime. This study seeks to strengthen the growing efforts to redress these conditions through a collective action by state and various fisher actors. In this focus, the research shares the contention that resolution of resource allocation dilemmas and responses to resource degradation rest on effective institutional co-ordination of public and private management. There has been a long history of fisher management in the state of Kerala. Of late, the mechanized sector has also joined hands with traditional fishers to strengthen efforts to manage marine fisheries. The private actors-fishermen groups, mechanized fisher groups, their associations and trade unions-work together and negotiate directly on many issues that confront resource health, sustainable livelihoods, sustainable profits from fishing operations and conflict resolution.

Unfortunately state bureaucracy responds slowly to these issues, enforces unilateral authoritarian decisions, excludes private actors from resource management, and hence fails to ensure the desired outcomes of management. Informal fisher negotiations with the state on the other hand deliver the socially accepted outcomes.

Technological developments introduced during mid nineteen fifties and the subsequent liberalization processes into the maritime economy of Kerala enhanced fish production and foreign exchange resulted in a diverse coexistence of artisanal and modern fishing methods and stakeholders who retained and pursued conflicting interests. Socio-political mobilizations that objected such lopsided developments resulted in state regulations that finally favoured collective actions of economically marginalized fishing communities.

Today, traditional fishermen who are the immediate subjects of serious livelihood crisis and mechanized boat owners who could not make reasonable profits to sustain their activities demand for better governance through local organizations. Artisanal fishers allege the mechanised counterpart for the entire environmental and socioeconomic crisis; the competitors counter these allegations and attribute the crisis to wrong fisheries policies, inadequate institutions and presence of industrial fleets. Major causes of confrontation between them still persist. However, interactions take place between apex level organizations, sometime directly but most often through the involvement of state. The state attempts to resolve the crisis by introducing a ban on mechanised fishing during the months of monsoon when the most targeted fishery is believed to breed. Although this measure has been largely accepted by artisanal fishers our study indicated that the economic and social costs of introducing such regulations exceeded the immediate gains of closing seasons and therefore better alternatives need to be developed.

The major reasons for the failure of state to evolve a dynamic system of institutions are it's over confidence on its own top-down bureaucratic management authority, underestimating community based initiatives, institutions and organisations in management and the lack of sufficient political brokerage to offer better governance. Analysing fisheries policies and formal legal systems, we noted that the federal system of management crafted formal legal systems at two levels which most often faced severe problems at the level of implementation. Moreover, no attempt has been made by the state to regulate entry of foreign fleets into Indian waters. The lack of effective institutions to regulate fishing efforts by artisanal, mechanised and foreign fleets continues to challenge the crafting of a management regime.

One of the major outcomes of this state failure has been the evolution and growth of non-state legal systems taking control over the regulatory processes in marine fisheries. The need for better regulations has been accepted by both parties and the thrust was to evolve collective action to organise sustainable fisheries that guarantee livelihoods for artisanal fishermen and profits for the existing industry. A number of informal organisations and political processes have contributed to the evolution of this cooperative social action in fisheries sector. These institutional processes are the product of multi-level "informal negotiated consultations and settlements" among state and various fisher actors. The crucial targets of management are the choice of technology for the contesting groups and the capacities and effectiveness of such choices to sustain livelihoods economic profits and conflict resolution. This was not an easy task as negotiations attempted integrated solutions than accepting any single solution that satisfies one group or the other.

As formal managers did not legitimise these non state legal systems and negotiated resolutions, which they consider as illegal solutions, fisher organizations and political parties in their capacity as law makers negotiate these disputes to derive temporary solutions. Examining the present level of community, industry, state interactions, we found that coastal fishermen along Cochin coastal zone of Kerala, India, have evolved strategies that not only regulate fisheries but also deliver sustainable livelihoods to participating fishermen.

As part of this restructuring, the traditional sick fishermen innovated and adapted appropriate technologies by reviving their traditional "socio-cultural" values. This restructuring, although illegal, empowered them to recapture their fishing territories surrendered to the mechanized sector and sustained livelihoods. The private industry on the other hand struggles to sustain profits mainly due to lack of incentives and state policy failures that open up coastal resources to international fishing fleets.

An alternate arrangement that network public and private agents through a political process has already evolved and commands management functions informally. This coalition has definite advantages for delivering sustainable solutions to livelihoods and resource health. State however, has to yield and tolerate these evolutionary proposals as this is the only channel through which local communities negotiate their immediate livelihoods from fishing today.

For artisanal fishermen, resorting to the politics of legal plurality that legitimises indigenous technological innovations and their choices and distribution of social product has been the only viable social arrangement to sustain livelihoods. In that sense legal pluralism is an inevitable pre-condition for good marine fisheries governance. Our study revealed that both state and private mechanised fishing industry gain from this partnership in the form of huge political benefits for the former and profits for the latter. The private sector has no option now but to participate in the negotiation process as the choice to abstain from the state-community partnership demoralises the sector and its commitments to evolve management regimes that sustain marine fisheries.

THE GLOBAL FISHERIES CRISIS

Records kept by the United Nations Food and Agriculture Organization (FAO) also indicate that of the world's 15 main fishing regions, four are depleted and nine are declining. This global fisheries crisis is primarily a result of overharvesting. The world's marine catch has increased more than four times in the past 40 years--from 18.5 million tons in 1952 to 89 million tons in 1989, but that growth is at great cost to the environment, and ultimately, perhaps to world food security.

Several decades of overfishing in most of the world's major fisheries has pushed many commercially important fish populations into steep declines Canada's northern cod collapse, for example. Catches are falling, despite the fact that expanding fleets are fishing harder, spending more time, effort and money than ever before in trying to maintain them. Some commercially important stocks are in such a critical state that all fishing has been shut down, or sharply curtailed. Hundreds of millions of people traditionally dependent on fishing for food and livelihoods face resource depletion, competition from industrial and distant water fleets, and loss of access to traditional marine food supplies.

The biological diversity of the world's oceans is threatend by the highly competitive race to catch enough fish to keep pace with rising international demand.

Rising demand and decreasing stocks have led to increased mechanization, over-capitalization of the industry and the buildup of excessive fishing fleets, particularly of the larger-scale fishing vessels that are responsible for the bulk of overfishing related problems worldwide. A modern 'factory' supertrawler can be longer than a football field and capable of catching and processing into various products up to 200 tons of fish daily. One of the world's biggest trawl nets could encircle more than a dozen "jumbo jet" Boeing 747 aircraft at its opening. Ships deploying such nets have a capture rate of about ten tons of fish per hour.

Since 1970, the world's fishing fleet has expanded twice as fast as world catches. The fishing fleet in China, the world's leading fish producer, is now around six times the size it was in 1979. As a result, excess fishing capacity has reached alarming proportions. There are about 3.5 million vessels boats currently fishing in the world's oceans. Intense competition between countries and rival fleets over access to fishing grounds has sparked numerous international disputes over fishing rights in recent years.

Fishing for Protein

Since 1989 the world~s annual catch has been dropping. Many argue that this decrease is an indication that fishing has oversteppped natures~s limists.

The FAO warns that the rising demand for fish and fish products, combined with shrinking global catches from declining stocks, will soon lead to the point where there will be a shortfall of fish for human consumption of more than 20 million tonnes each year. Average annual consumption of fish caught in marine and inland waters could fall from 10.2 kilograms per person in 1993 to somewhere between 5.1 and 7.6 kilograms by 2050. This threatens the one billion people, mostly in developing countries, who rely on fish as a principle source of protein.

Wasteful and Destructive Production

The increased fishing pressure and the competition amongst fishing nations and their fleets severely stresses fish stocks and the marine environment. The widespread use of unselective fishing gear and indiscriminate practices result in one-quarter of the all the fish brought on board fishing vessels each year being discarded, usually dead or dying, back to the sea.

Commercial fishing vessels throw back on average about 27 million tons of unwanted fish annually. That amounts to about half of all the fish caught from the oceans each year that are consumed directly by humans. Along with these, millions of other marine animals are being incidentally captured and killed in fishing operations.

Destructive Fishing Gear

Some fishing gear is particularly deadly for certain animals in some situations. Longline fishing boats kill many tens of thousands of albatross each year in southern hemisphere oceans. Driftnets indiscriminately killed millions of marine creatures, while targetting for just one or two commercially valuable species. Marine mammals are frequently killed in great numbers in trawls, set nets and purse seine nets. In addition there is severe damage caused by fishing operations that use destructive gear and fishing practices, like bottom trawling, that physically disturbs marine habitats such as the ocean floor, sea grass beds or coral reefs.

The Politics of Overfishing

Today, there isn't a fishing region in the world that does not suffer from fisheries management decisions designed to satisfy short-term economic or political objectives (or both) rather than protecting the marine environment and conserving fish populations. The bulk of the problems stem from twenty fishing countries whose fleets land 80 percent of the total marine catch worldwide.

1. China	2. Peru
3. Japan	4. Chile
5. USA	6. Russian Fed.
7. Thailand	8. Indonesia
9. Korea, Rep.	10. Norway
11. India	12. Iceland
13. Philippines	14. Korea, DPR
15. Denmark	16. Spain
17. Taiwan, Prov China	18. Canada
19. Mexico	20. Vietnam

Commercial fishing in most of these countries has been very poorly managed. Even in a few countries where relatively advanced fisheries management systems have been in place for many years they have, almost without exception, failed to control the conditions and stem the abuses that lead to overfishing and destructive environmental impacts. Indeed, in many countries, governments have played an important part in fueling the expansion of excessive fishing capacity and overexploitation by providing lucrative subsidies, taxpayer funded handouts. On a global scale, these destructive subsidies run up to $50 billion a year.

The chronic failures of fisheries management in European waters of the northeast Atlantic and North Sea are particularly noteworthy since this area has probably the longest standing and greatest single concentration of fisheries research and management institutions in world. Yet, Europe's fish stocks are plagued by overfishing, and massive excess fishing capacity.

The member countries of the European Union regularly ignore scientific advice when setting their annual catch quotas: when scientists from the International Council for the Exploration of the Seas recommended a 40 percent cut in the 1995 hake catch to protect the stocks, EU fisheries ministers agreed to a mere five percent cut. In the EU, as elsewhere, it seems that a fishery must be proven to be on its death bed before any remedial action is taken.

In the United States a similar picture of failed fisheries management prevails. According to the National Marine Fisheries Service (NMFS), the federal agency overseeing U.S. fisheries, 80 percent of the known commercially valuable fish populations (122 aquatic species) are currently either fully or overexploited; 67 species of these, among them bluefin tuna, swordfish, oysters, hard-shell and soft-shell clams and red snapper--are overexploited. A further 79 species continue to be exploited even though the status of their populations is unknown. Little wonder why this is so. The country's eight regional fisheries management councils are primarily composed of those interests who benefit directly from increased catches: commercial fishers and other industry user groups.

It does say much for fisheries management regimes elsewhere in the world, if the European Union and the United States are considered to be amongst the leaders in the field of fisheries conservation and management.

Quite simply, nature's limits have been breached by too many fishing vessels catching too many fish, very often in wasteful and destructive ways, and it cannot allowed to continue if the oceans and the human communities around the world that depend on them are to survive. Governments and industry are responible for ensuring swift action to bring about urgently needed reforms, including substantial fleet reductions and substantial cutbacks in allowable catches, the elimination of destructive subsidies, major reductions in bycatch and waste, and more stringent environmental controls over fishing operations. The full utilization of available fish stocks and profit maximization for industry have been the key goals of short-sighted fisheries development, while protection for the environment has taken a back seat. This has proven to be the formula for disaster in fishery after fishery, the world over, with the disastrous consequences for marine ecosystems and humanity already plainly visible around the world.

GLOBAL TRADE AND CONSUMER CHOICES: CORAL REEFS IN CRISIS

Overview

Coral reefs around the world are in crisis. Over 25% have already died or are severely damaged, and another 30% are seriously threatened and may die--from global warming, overfishing, destructive fishing practices, and unsound coastal development. Contributing to the problem, many coral reef animals and products are collected globally for commercial purposes, including food fish, the marine aquarium trade, live food fish markets, curios and knick-knacks, jewelry, and traditional medicines. Global trade is leading to overexploitation of reef animals and the use of fishing practices that destroy the reefs. Recent surveys of reefs worldwide found that many species of high commercial value were absent or present in very low numbers, in almost all the reefs surveyed. Results suggest that almost all coral reefs have been affected by overfishing, and that there may be no pristine reefs left in the world.

The United States is the number one consumer of live corals and fish for the marine aquarium trade, and of coral skeletons and other dried animals for the curio and jewelry markets. American consumers are unknowingly contributing to the decline and degradation of coral reefs. As a major importer and world leader in both trade and coral reef conservation efforts, though, the U.S. can play a critical role in shaping conservation strategies, consumer awareness, and international trade policies.

The symposium, Global Trade and Consumer Choices: Coral Reefs in Crisis, held at the 2001 American Association for the Advancement of Science Annual Meeting and Science Innovation Exposition in San Francisco, California, featured five experts who spoke on the global trade in coral reef species and coral reef conservation. They analyzed the causes and consequences of global trade in coral reef animals, the role of the U.S. as a major importer, and presented solutions to reduce the negative impacts of trade on these unique ecosystems while promoting long-term benefits to local communities through sustainable resource use. This report presents their papers, as well as the following overview of their main points.

Franklin Moore and Barbara Best of the U.S. Agency for International Development note that "coral reefs are invaluable resources to local communities around the world, serving as sources of food, jobs and livelihoods, and as coastal protection." By one estimate, coral reefs provide economic goods and ecosystem services worth about $375 billion each year to millions of people. "However," observe Moore and Best, "international trade is driving overfishing and destructive fishing practices, such as the use of cyanide to collect live reef fish. The unsustainable and destructive use of these precious resources jeopardizes the potential of coral reefs to sustain local communities and future generations."

Roger McManus, former President of the Center for Marine Conservation and Senior Advisor for Oceans in the U.S. Department of the Interior, notes that the trade in coral reef species is in actuality trade in an entire ecosystem. McManus asks if this trade, though, is consequential in light of other threats posed by global change and pollution. Fishing and collecting of other coral reef species are probably not sustainable in most cases, are the most significant threat to many targeted species, and in many areas are a significant threat to the overall health of reef ecosystems. McManus adds "it is culturally and politically important to address all threats to natural resources to ensure equitable treatment."

McManus also argues that the U.S. should not allow the World Trade Organization to influence its actions in the global trade arena. "The U.S. should exercise its authority as an international leader in both trade and environmental conservation, and address its own trade policies on coral reef animals." He suggests that the Convention on the International Trade in Endangered Species of Wild Fauna and Flora (CITES) "was not designed to be implemented as a comprehensive wildlife management tool at this scope of [thc coral reef] trade." McManus recommends the creation of more marine protected areas with no-take zones, as well as a phase-out of U.S. imports of wild coral reef species except where the trade can fulfill criteria for sustainability more stringent than those now in use under CITES.

Charles Birkeland, from the University of Hawaii's Hawaii Cooperative Fishery Research Unit, noting the diversity of coral reef ecosystems, the life history adaptations of reef species, and the close knit intricacies of species interactions, cautions that coral reefs are inherently more vulnerable to overuse than are other marine ecosystems. If the coral reef trade is to be sustainable, reefs must be managed using an ecosystem-based approach. However, the fisheries management laws of the U.S. constrain the use of best management practices, such as an ecosystem-based approach, and are particularly ill suited to promote sustainable management of coral reefs.

The Coral Reef Ecosystem Fishery Management Plan for the western Pacific, though, is charting new territory as the first ecosystem-based fishery management plan for U.S. waters. It adopts a precautionary approach that shifts the burden of proof through a permitting system, applies adaptive management through continuous use of new information, creates marine protected areas with no-take zones, and requires "insurance" against unforeseen ecosystem impacts.

There is already strong international concern that some coral reef species are threatened or may become threatened through trade. Those species are listed under CITES, and include 2000 species of hard (stony) corals, black coral, giant clams, Queen conch, and sea turtles. However, most of the coral reef animals in trade are not covered under CITES.

Susan Lieberman, former Chief of the U.S. Fish and Wildlife Service (USFWS) Division of Scientific Authority and currently Director of the Species Programme of WWF International,and John Field, also with USFWS, discuss the potential for international trade agreements, such as CITES, to address environmental concerns. They note that "under CITES, exporting countries are required to conduct scientific determinations that this trade is sustainable, and will not jeopardize species or their ecosystems." However, this ideal is often not fulfilled in reality. While the primary responsibility for ensuring the sustainability of harvests and exports under CITES rests with exporting countries, Lieberman and Field suggest that "as an importing country, [the U.S.] also must share some of the responsibility, whether through cooperative bilateral and multilateral efforts or regulatory actions." They add that it is "vital to couple those strategies with increasing public awareness... An informed consumer is powerful 'Äì and can guide industry best management practices, government policies, and of course Congressional interest and action."

Vikki Spruill and Lisa Dropkin, of SeaWeb, though, observe that "Americans have little knowledge of the real threats to coral reefs and the life they support." Surveys show that Americans believe pollution is the greatest ocean threat, and that most consider coral reef conservation a second tier concern. But, there is support for coral reef protection and there is growing momentum to use consumer choice to drive ocean conservation. Spruill and Dropkin note that "Americans often express their conservation values by what they purchase. We need to make Americans aware of how consumers can contribute to reef conservation by buying sustainably caught reef animals." They add that informal data suggests aquarium hobbyists want an industry based on sustainability and are willing to modify their purchasing behaviors to support reef conservation.

Clearly, there is an urgent need to address the trade threat to coral reefs. Dire as the picture may seem, the papers presented here point to a way forwards. Potential remedies are known. Some are already being tested 'Äì like coral reef reserves with no-take zones, and consumer awareness campaigns with likely application to reef species. To achieve an equitable and sustainable solution will require that all must be pursued, engaging communities, exporters and importers, governments, non-governmental organizations, consumers, scientists, and international institutions.

CORAL REEF CRISIS: CAUSES AND CONSEQUENCES

This paper presents an overview of the value of coral reef ecosystems to developing countries, the impacts of international trade on coral reefs and local communities, and the role of the U.S. as a major consumer nation and driving factor in the international trade of corals and coral reef species.

Coral Reefs are Invaluable Coastal Ecosystems

By any measure, coral reefs are among the most diverse and valuable ecosystems on earth. Coral reefs occur in over 100 countries, most of them developing countries without the capacity or financial resources to adequately manage these vital resources. Reefs support at least a million described species of animals and plants, and another 8 million coral reef species are estimated to be as yet undiscovered.

According to one estimate, coral reefs provide goods and services worth about $375 billion each year-a staggering figure for an ecosystem which covers less than one percent of the earth's surface. Reef systems provide economic and environmental services to millions of people as shoreline protection from waves and storms, as places for recreation and tourism, and as sources of food, pharmaceuticals, livelihoods, and revenues.

In developing countries, coral reefs contribute about one-quarter of the total fish catch, providing food to an estimated one billion people in Asia alone. Globally, half a billion people are estimated to live within 100 kilometers of a coral reef and benefit from its production and protection. In light of expected climate change and associated sea level rises, coral reefs can offer a natural, self-building and self-repairing breakwater against wave and storm damage. These extremely valuable ecosystems constitute the economic base and future hope for sustained development in many countries, particularly small island nations.

Coral Reefs in Crisis

A recent report from the Global Coral Reef Monitoring Network estimates that 25% of the world's reefs are already gone or severely damaged and that another third are degraded and threatened. This coral reef crisis is happening here at home in the U.S. and in far away places, in some of the most remote areas of the world.

Coral reefs are in serious trouble worldwide from a powerful combination of stresses that are threatening their survival, including:

- overexploitation of resources for subsistence and commercial fishing;
- destructive fishing practices that degrade and destroy the habitat itself;
- increasing coastal populations, which are expected to double in the next 50 years;
- poor land use practices and runoff of pollutants, sediments and nutrients;
- disease outbreaks, which may be associated with poor water quality and pollutants;
- coral bleaching, associated with increasing seawater temperatures and global change; and
- removal of coastal mangrove forests.

These direct and indirect human activities pose significant threats to coral reef ecosystems, and the human populations that depend on them, particularly small island developing countries. For example:

- In northern Jamaica, it is estimated that almost all of the reefs are dead or severely degraded from overfishing and coastal runoff. Fish stocks have declined to a point where local fishers are now straining fish larvae out of the sea for fish soup.
- In the Philippines, degraded reefs and fish populations have led to an 18% decrease in the amount of protein in the average diet.
- Human impacts are also occurring on U.S. reefs, oftentimes for use as luxury items. For example, in Hawaii at Honaunau, the top ten aquarium fish species have decreased by 59% over the last 20 years, and at Kona the most popular aquarium fish show declines in abundance from 38 to 57%.

Even under ideal conditions, it would take more than a lifetime for some reefs to recover. We can no longer continue to take coral reefs or mangrove forests for granted, or to assume that they can support unlimited resource use or unmanaged global trade.

Trade, Mangrove Forests, and Coral Reefs

While coral bleaching may be one of the largest threats facing coral reefs, international trade is having significant impacts on even the most remote and pristine reefs Recent surveys of reefs worldwide found that many species of high commercial value were absent, or present in very low numbers, in almost all the reefs surveyed (Hodgson 1999). Results suggest that almost all coral reefs have been affected by overfishing, and that there may be no pristine reefs left in the world.

International trade is also posing significant threats to mangrove forests, another critical coastal ecosystem that is intimately connected to coral reefs. Mangrove forests serve as important nurseries for many reef species. They help to maintain coastal water quality by reducing the run-off of sediments, pollutants and excess nutrients from the land. Nutrients and energy flow between these habitats as species move between them.

In some areas of the world, the major loss of mangrove forests is due to the construction of shrimp mariculture ponds for the world market. The cheap shrimp we consume here in the U.S. comes with enormous ecological and social costs for the local communities where mariculture ponds are inappropriately sited and intensively farmed.

Trade Drives Destructive Fishing Practices

How does the international trade in wild coral reef animals and products more directly impact reefs? Primarily through overfishing and the use of destructive fishing practices. Live fish for both the food trade and marine ornamental trade are often caught with the use of cyanide or other poison, which temporarily stuns the fish for easy collection. Cyanide use is a serious threat to some of the world's richest coral reefs, as the cyanide kills corals and many other coral reef organisms. The lucrative and unregulated international trade in reef fishes drives the use of cyanide. It is estimated that since the 1960's, more than one million kilograms of cyanide has been squirted onto Philippine reefs alone, and the practice has spread throughout East Asia and the Indo-Pacific (Bryant et al., 1998).

Various explosives, such as dynamite and homemade bombs, are also used to kill fish for easy collection, but at an enormous cost to the reef which is reduced to rubble. In Komodo National Park in Indonesia, about half of the reefs have already been destroyed through the use of explosives, forming beds of coral rubble that can extend several football fields in length. While the use of explosives to collect dead fish is usually for domestic trade, some of the fish that are only stunned will enter the international trade stream.

International trade is also driving the removal of the calcareous skeleton or base of the reef itself; reef skeletons are sold as "live rock" for marine aquaria. This base is the resulting accumulation of coral skeletons over tens to hundreds and thousands of years. Living coral, which constitutes the essential reef habitat for a myriad of species, is also collected and shipped live for marine aquaria, or killed and dried for the curio and shell trade.

Trade Drives Overfishing and Removal of Targeted Groups

In addition to destructive practices, international trade is driving overfishing and the selected removal of key groups from coral reefs. Major groups targeted for trade are:

- groupers and wrasses for the live food fish trade;
- dead fish and invertebrates for food, medicinal products, and ornamentals including sharks, sea cucumbers, sea stars, mollusks and sea horses;
- live fish, coral and other invertebrates for marine aquaria and the ornamental hobby; and
- "live rock" or the calcareous base of the reef for marine aquaria.

The marine ornamental trade for the pet industry often targets rare fish and coral species, as these can fetch the highest prices. The trade is also targeting large-polyped corals, which tend to be the slowest growing and the least common. By targeting the large groupers and wrasses, the live food fish trade removes key species from these ecosystems, thus altering their dynamics. The loss of some is comparable to the loss of major predators from terrestrial ecosystems.

Other fishes feed on algae, and thus play an important role in ensuring that corals are not overgrown by more rapidly growing algae. The removal of coral for the marine aquarium trade and for use as curios and knickknacks, and the removal of the "live rock" base, reduces the essential reef habitat.

There are strong economic incentives associated with this international trade. The live food fish trade through Hong Kong alone is estimated to have a retail value of about one billion dollars a year. Some species of fish, selected live from a restaurant tank, can sell for almost $300 per plate. The global retail of marine ornamental fishes and aquarium hobby supplies is estimated at $500 million. Last year, for example, a pair of rare fish sold for over $5,000 each. Over 1000 different species of coral reef animals are now traded for marine aquaria.

The impacts from international trade are quite different from other more chronic causes of reef degradation, as these impacts are felt even in the most remote, pristine reefs. The use of destructive fishing practices, such as the use of cyanide, is spreading throughout the Indo-Pacific as fishing boats venture farther to find new unexploited fishing grounds

There is already strong international concern that some coral reef species are threatened or may become threatened through trade. Those species are listed under the Convention on the International Trade in Endangered Species of Wild Fauna and Flora (CITES), and include 2000 species of hard (stony) corals, black coral, giant clams, Queen conch, and sea turtles.

Consequences of Destructive and Unsustainable Fishing Practices on Local Human Communities

International trade is driving destructive fishing practices and unsustainable harvests from coral reef ecosystems, reducing the value of coral reefs to local communities and prospects for long term sustainable use. In some areas, depletion of stocks and the destruction of the reefs are threatening peoples'Äô food security. This international trade is a highly mobile trade; as stocks are depleted in one area or country, the trade moves on to other areas, often spreading the use of destructive fishing practices. Thus, the nature of the trade provides few incentives for long-term sustainable use by a community, and few benefits are channeled to the local communities.

The use of poisons and hooka gear can have serious consequences for the collectors themselves. Cyanide fishing poses human health risks to the fishers through exposure to the poison. A hooka rig is a low-tech approach to scuba diving that involves a compressor on the boat that pushes air down long tubes to divers below. Divers can spend many long hours under water collecting with hooka rigs. Unsafe diving practices by untrained divers can lead to the diver's "bends" and result in joint disease and even paralysis and death. Each week, several divers who have contracted the bends are taken by fishing boats in Honduras for treatment in the local diving decompression chamber. These divers have been collecting spiny lobsters to supply the growing U.S. appetite for seafood. Similar reports of injuries to divers come from South East Asian countries where hooka rigs are used for collecting marine ornamental fish and live food fish.

U.S. Role in International Trade

In 1998, in response to the coral reef crisis, the Executive Order for the Protection of Coral Reefs was signed. The Order created the U.S. Coral Reef Task Force as a way of coordinating federal and state efforts, and charges federal agencies with the conservation and sustainable use of reef resources both domestically and worldwide. The Task Force was also asked to analyze and address the U.S. role in the international trade of coral and coral reef species.

The results of the trade analysis reveal that while live reef fish for the food fish market primarily go to Asian markets, the U.S. is the number one consumer of live coral and marine fishes for the aquarium trade and of coral skeletons and precious corals for curios and jewelry. Inadvertently, American consumers are contributing to the worldwide decline and degradation of reefs.

A closer examination of the U.S. trade reveals that the U.S. was consistently the largest importer of live coral during the 1990s, importing over 80% of the live coral and 95% of the live "rock" or reef base. Ironically, the U.S. prohibits the collection of coral and live rock in its own waters as they are considered essential fish habitats.

In addition to coral, the United States imports nearly half (eight million) of the total worldwide trade in aquarium fishes (15-20 million/year). Many of the fish imported for the marine aquarium market in the U.S. are captured with the use of cyanide and other poisons, which kills non-target animals and the coral reef itself. Sustainability concerns will only increase with the growing international trade. The international trade in coral and live rock to supply the aquarium trade has increased at a rate of 12 to 30% per year since 1990.

The U.S. Role in Addressing the Trade Threat

The U.S. is part of the problem. The U.S. needs to be part of the answer. As a major consumer and importer.

of coral reef organisms, a major player in the world trade arena, and a leader in coral reef conservation efforts, the U.S. has a critical responsibility to not only address the degradation and loss of coral reef ecosystems worldwide, but to also encourage more responsible trade. As consumers, the U.S. should discourage the use of destructive or unsustainable collection practices that may jeopardize the future potential of coral reefs to sustain the local communities who depend on them for food and livelihoods. Rather, we should reward and encourage responsible use of these precious resources, and shift the burden of proof of sustainable use, for commercial and recreational purposes, to the users.

We need to emphasize community-based management of coral reef resources so that people living on and around coral reefs may share in the profits from coral reef activities. Other exploiters, whose primary interest is in making money without sharing benefits with local communities, should not be allowed to profit from these precious resources.

Oftentimes, local communities or national fisheries departments lack the capacity to sustainably manage reef resources, or to resist the short-term, high gain, economic incentives associated with the live food fish and marine ornamental trades. The U.S. Agency for International Development (USAID) is the principal agency of the U.S. Government responsible for building capacity in sustainable resource use in developing countries. USAID is presently working with local communities and national governments in about 20 countries to assist them in conserving and managing their coral reef and coastal resources through capacity building for integrated coastal management, better land-use practices, sustainable fisheries management, and marine protected areas.

There are also immediate actions available to the U.S. public in terms of awareness and individual consumer choice. There is an urgent need to develop positive trade regimes so that only products from reefs under sustainable management plans are allowed into or out of the U.S., to ensure that consumer demand by Americans is not contributing to the decline and degradation of coral reefs.

We must change our view of how we treasure and value natural resources. For example, last month, the U.S. adopted new trade measures covering the import of antiquities from Italy into the U.S.; all antiquities from Italy must now be accompanied by documentation and certification has to how they were collected and where they are from, to ensure that they are from legitimate sources. We must take a similar approach to natural resources. The U.S. government is promoting the idea among other nations within the Asia Pacific Economic Cooperation forum that consuming nations must bear some of the responsibility for their imports, and it is considering this approach for reef resources.

The U.S. could play a significant role by helping reward responsible practices, creating market incentives for responsible behaviors, and requiring certification of non-destructive collection practices and demonstration of sustainable collection of coral reef species. In this way, government, consumers, hobbyists and industry members can work together to ensure a responsible trade.

U.S. EFFORTS TO PROTECT DOMESTIC AND INTERNATIONAL CORAL REEFS: TRADE IN THE LARGER CONTEXT

Trade in an Ecosystem

Well over a year ago I attended one of those Georgetown dinners in which policy makers and those that would influence them meet to come to political terms. The order of the evening was to convince environmental organization leaders to support Clinton Administration policies regarding free trade agreements and the World Trade Organization (WTO).

A focus of the discussion was the trade in coral reefs. I say coral reefs, not coral reef organisms, because as recent trade studies have documented, the trade is in dead and live coral, other coral reef invertebrates and associated algae, and fishes. This is literally trade in an ecosystem; the majority of the trade is in its underlying structure. (There is a conversation evolving referring to "biotic" reefs recognizing that a majority of reef components are not living corals, but including algae and other "reef" building invertebrates.) In any case, the largest importer by far is the U.S., primarily for the aquarium trade. The U.S. is a major importer, too, of sea horses and Queen conchs.

Is this trade consequential in light of global change, reefs being mined for building material, or otherwise destroyed by pollution? The answer is probably mixed. Global change is the major threat to reefs worldwide, and mining and pollution probably are the next biggest threats accounting for the largest losses. Nevertheless, fishing and harvest for the aquarium trade are probably not sustainable in almost all circumstances, and in many areas may be the most significant threat to coral reefs. In cases where the trade is targeted at rare species, the trade may be a threat to their continued survival. Moreover, it is culturally and politically important to address all threats to natural resources to ensure equitable treatment. However one gives weight to the relative threats, the destruction of coral reefs by human action is clearly ongoing.

All of this continues despite the fact that:

- tourism is emerging as the number one world industry, and that coral reefs have enormous economic value as a sustainable basis of tourism for many countries, for fisheries, and for other purposes;
- many coral species are recognized as threatened by trade, or potentially so, and are protected by the Convention on International Trade in Endangered Species of Wild Fauna and Flora (CITES);
- there is clear acknowledgement that harvest of coral reef organisms is a significant contributing factor to the degradation of coral reefs;
- the U.S., recognizing the threat of trade to coral reefs, largely restricts the extraction of corals and coral reef organisms from most of its own federal, state and territorial waters; and
- many of the fishes taken for export for food and for the aquarium trade are captured using cyanide and other poisons, or other methods destructive to their habitats. The result is more often than not consumer fraud as diners and aquarium fanciers frequently buy fish that are in the process of dying from the poisons used to capture them. Many of the divers conducting the captures are exposed to the poisons, and are not trained divers, increasing incidents of death and injury.

The World Trade Organization and Environmental Controls

During that evening in Georgetown, I noted that the Clinton Administration would do nothing to curtail its role in this destruction. I reasoned that the Administration simply would not take on the World Trade Organization (WTO) and its policies supporting free trade over any other concern. State Department officials assured me that I was wrong. But, sadly, I was right, the U.S. would effectively do nothing to stem imports of coral reefs into our county and the international trade continues to grow virtually unabated. We have instead essentially limited our efforts to assisting exporting nations to care for their coral reefs, to education, and to appealing for voluntary restraint by consumers. Why can't the world's only remaining superpower curtail its domestic commercial imports of coral reef products when it is clear that the trade is unsustainable, destroying reefs, and undermining the present and future economies of developing countries?

Because the World Trade Organization won't let us. If Congress passed a law controlling trade in imports of coral reef products, one or more of the exporting countries would complain to the WTO that we are inappropriately interfering with free trade. The burden would be on the U.S. to prove to the WTO's satisfaction that our actions were justified. This may seem fair enough, except that the WTO has upheld every challenge to environmental controls, and the burden of proof is always on those seeking to help conserve and protect the environment and our natural resources.

If a country does not agree to curtail its "unfair" efforts to protect the environment, then the court provides for the offending country to suffer financial penalties. If the U.S. won't stand up to offended nations, consider what a small developing country would do in the face of threatening complaints from another nation bent on profiting from the elimination of offending pollution controls or wildlife protections?

WTO's unbridled advocacy of trade no matter what its social and environmental costs is unacceptable if we are going to foster the kind of natural resource stewardship that ironically will make sustainable economies possible.

That is why worldwide opposition to the WTO is rising and won't go away.

Of course it is not necessary to open all environmental controls to attack to prevent some countries from trying to promote inappropriate protectionism. A solution to this problem must be forged if we are to continue the benefits of expanding world free trade, and protect our planet's environment and its wildlife.

The highest priority for the new U.S. Trade Representative, Robert Zoellick, is to secure "fast-track authority" from the Congress for the President to negotiate new trade agreements. He has criticized the arguably modest efforts by the Clinton Administration to ensure trade agreements do not undermine environmental protection. During his confirmation hearing, it was clear that Congress will be facing a contentious battle on how fast-track legislation will ensure needed environmental protection. One way to resolve conflict is to find equitable solutions.

Can Trade in Coral Reefs be Sustainable? The Limitations of CITES

Through Executive Order no. 13089 for the Protection of Coral Reefs, President Clinton directed the Secretary of State and the Administrator of the Agency for International Development, in cooperation with other members of the Coral Reef Task Force to assess the role of the U.S. in international trade in coral reef species, and to implement strategies and promote conservation worldwide to protect coral reefs. The report from the Trade Subgroup of the International Working Group to the U.S. Coral Reef Task Force helps to respond to this directive.

The conclusions presented above about the threats of trade to coral reef ecosystems echo those of the Working

Group. The report noted an accumulating suite of references to coral reefs providing economic and environmental services worth hundreds of millions of dollars a year. The report suggests that almost all reefs have been affected by over-harvesting, and that there may be no pristine reefs left in the world. In addition the Working Group made numerous recommendations on how to address the problem, including establishing new authorities to reduce the adverse impacts of the trade.

There is a fundamental assumption in these recommendations that it is possible to have sustainable, commercial exploitation of coral reef ecosystems. Many coral reef species are slow growing and slow to mature. Reefs are generally characterized as ecosystems with large numbers of species with relatively few numbers of individuals. Whether one measures sustainability by sustained biomass production, biomass remaining from exploitation, or stability in biodiversity measures, many observers of coral reef ecology are skeptical that such exploitation can be conducted in a relatively benign manner. More likely it is through a series of no-take reserves in marine protected areas, and rotating exploitation in sacrifice zones, that the impacts of the exploitation can be managed. Such management regimes in Australia have been held out as successful examples of this approach. Whether such a regime could be held to be really sustainable would presumably be dependent on whether some steady state over some meaningful time frame could be documented.

Nevertheless, while more research is needed to ascertain the impacts of all threats to coral reef ecosystems, and to demonstrate sustainability of their use, the burden of proof now generally lies with those who would constrain exploitation. Current public process assumes that those who profit from these resources are conducting their activities in a sustainable manner notwithstanding our knowledge of coral reef biology and data on harvest methods, quantities, and impacts that belie such an assumption.

That is not to suggest demonstrating confidence in sustainability would be straightforward. Thousands of species are involved in the trade, and the general tenor of the Working Group report is to seek sustainability on a species by species basis. No doubt, particularly with respect to live rock, many of the species in trade have not been described, the taxonomy of others is under debate or difficult to ascertain, and for most their biology is so imperfectly understood that our ability to suggest any level of exploitation as sustainable is speculative at best. Coupled with the fact that most organisms in this trade are identified in public documents at best to genera or even higher taxa, the conversation that meaningful decisions about the sustainability of harvest or trade is happening or could be so is highly questionable.

Imagine for a moment you are an official in Fiji in charge of issuing export permits under CITES. For stony corals under Appendix II you are charged with ensuring that the species to be exported were taken legally and in a manner that was "not detrimental to the survival of the species." There are 2000 corals listed under CITES, and you are presented with an export application for "live rock," including some species controlled by CITES, others controlled by other domestic law, and some for which there is no domestic or international law for their management. What do you do? What is your decision? How did you come to it? How do you decide this shipment and the ones before it and after it can be considered sustainable uses of the resource?

Keep in mind that the major substantive CITES standard for exports of Appendix II species is that the export is "not detrimental to the survival of the species". There are many people involved in this debate that view this standard as equivalent to a requirement for sustainability, but a clear reading suggests the CITES standard relates to probable extinction rather than a capacity to maintain commercial exploitation 'Äì arguably two very distinct standards. As the Fijian official you are lucky with regard to this standard. All you really have to suggest to allow the shipment is that it is not likely to be detrimental to the survival of any species it contains.

However, the treaty also suggests that exports should be limited to maintain the species throughout its range at a level consistent with its role in the ecosystem. Whether this is a legal standard for issuance of the permit that is enforceable is unclear. Whether "consistent with its role in the ecosystem" is a standard equivalent to "sustainable" is also uncertain. Sustainability, defined or not, is not a standard of international trade law, and for those purposes you are free, as that Fijian official, for all intents and purposes to ignore the question.

Imagine now that you are a border official in the U.S., and you receive a shipment of stony corals from Indonesia with an official permit attesting that the material was legally obtained and taken in a manner not detrimental to the survival of the species. You have several problems. You may not have a clue what species are really in the container, and if you did, how many are controlled by CITES or the U.S. Endangered Species Act. Under current U.S. policy, the pressure is to accept the shipment unless you challenge the authenticity of the export permit. You could also question the underlying assertions concerning whether the specimens were legally obtained or in the judgment of the Indonesian officials whether the shipment was detrimental to the survival of the species included. You are likely not in a strong position to do this. In the experience of CITES implementation, you are also on very thin ice with respect to the "role in the ecosystem" standard. Hence, you approve the import and waive the shipment through. These two examples suggest that CITES simply was not designed to be implemented as a comprehensive wildlife management tool at this scope of trade.

Recommendations

The International Working Group of the Coral Reef Task force identifies a series of actions being taken by U.S. agencies and makes recommendations that inherently make sense and are or would no doubt be beneficial, including research, education, better enforcement of existing domestic and international controls, and enactment of needed, restrictive legislation. The legislation did not happen, in part because some in the conservation community did not want interference with their "sustainable" coral reef projects. Nevertheless, enforcement of existing law is not likely to alleviate significantly the harmful impacts of the trade, except in some highly managed habitats

I would emphasize and add to those recommendations the following with the view to strengthen their potential impact:

1. Establish more coral reef marine protected areas with no-take reserves. This is undoubtedly the most efficient means to conserve the abundance and diversity of coral reefs. The U.S. has demonstrated leadership in this regard with the designation of the Northwestern Hawaiian Islands Coral Reef Reserve, and Palymra Island and Kingman Reef National Wildlife Refuges.
2. United States leadership has been less admirable in trying to resolve the conflicts between promoting

free trade and the conservation of natural resources. The U.S. government needs to be a leader in trying to find solutions.

3. The U.S. should phase out imports of wild coral reef species except in those specific cases where operations providing such commodities are operating in a manner demonstrated to fulfill some kind of numerically objective definition of sustainability. The current means and vision for controlling the trade are so disproportionate to its immensity as to render them largely ineffectual.
4. The U.S. government should reassess its role in dealing with global change, through its domestic management policies and international assistance, recognizing the impacts such change will have on coral reefs. One of the many consequences of sea level rise and global warming, will be the decline of coral reefs throughout much of their range.
5. All governments concerned with marine conservation should take efforts to reduce and discourage mining of coral reefs and to prevent otherwise their physical modification and pollution.

The 21st century will be the most important in the existence of coral reefs over the past 5,000 to 10,000 years. In this century, we will make important decisions either intentionally or by indecision on whether we want to conserve a majority of coral reef biological diversity and how many coral reefs we want to see survive into the next century.

With regard to the coral reef trade, what does it say about us as a species, when we continence destructive trade in one of our planet's richest and sensitive ecosystems? A trade that is argued excusable because it damages coral reefs less than other threats.

A trade that directly and predictably threatens the health and safety of our fellow humans. A trade that captures and holds other organisms for our aesthetic enjoyment when we know their lives are essentially treated as consumables being acquired largely by poisoning their bodies and maintained without expectation of their survival. How can we experience beauty and pleasure in such a process? What do we consider potentially sustainable about this exploitation?

GLOBAL SOLUTIONS TO GLOBAL TRADE IMPACTS?

Introduction

Coral reefs are some of the most productive and diverse ecosystems on Earth. Although they occupy less than 0.25 percent of the marine environment, reefs support more than a quarter of all known fish species (Bryant et al., 1998). Coral reefs are an example of a highly complex and interconnected ecosystem. The rich complexity found in the reef ecosystem is a result of the evolution of many symbiotic associations which couple photosynthesis with processes that increase the retention of limited nutrients within the system. Despite their high gross productivity, reefs rely on a fairly closed nutrient cycle that yields little surplus production for extraction. They serve as critical habitat for numerous tropical species, provide storm and wave protection for coastal communities, and yield renewable resources for human populations around the globe. Yet these same human populations, often in areas of rapid development and growth, are placing increasing stress on these vulnerable ecosystems, due to various economic and social pressures.

Coral reefs now face a multitude of anthropogenic stressors. According to a 1998 World Resources Institute report, nearly 58% of the world's reefs are at risk from a range of human activities 'Äì global warming, marine pollution, coastal development, destructive fishing practices, overexploitation of resources, disease, runoff from improper land-use practices 'Äì and many have been degraded beyond recovery (Bryant et al., 1998). The study concluded that overexploitation of resources, destructive fishing practices, and coastal development pose the greatest potential risks. The international trade in coral, reef fish, live rock, and other reef organisms also contribute significantly to the decline and degradation of reefs.

These stressors have a cumulative and synergistic effect on reef health, and often occur in countries that lack the institutional or financial capacity or political will to mitigate them. Thus, it falls upon the international community to continue to mobilize funding, foster cooperation, monitor trade, and provide expertise to coral range countries if the reefs are to recover and thrive in the 21st century. This paper focuses on the role of international trade, and multilateral and domestic efforts to regulate and control that trade.

The Role of International Trade

Coral reef species are traded as a variety of products in numerous international markets, including live reef food fish, traditional food fish, curios, traditional medicines, live marine ornamentals, coral and live rock for aquaria, limestone production, and construction materials. Each of these arenas has its own set of issues, management regimes, and industries but all are subjected to some degree of international control or scrutiny. The impact of international trade on coral reefs relative to other stressors (pollution, bleaching events, etc.) is not well quantified, but the fragility of reef ecosystems means that even secondary threats cannot be ignored. In fact, the partners within the International Coral Reef Initiative (ICRI) adopted a resolution in the fall of 1999 recognizing that "international trade in corals and coral reef species is contributing to the stresses on these systems."

The best available information on trade in coral species listed under the Convention on the International Trade in Endangered Species of Wild Fauna and Flora (CITES) is a recent study by The World Conservation Monitoring Center. Analysis of CITES records for black corals (from 1982-1997) and stony corals (from 1985-1997) revealed that 70 nations imported a total of 19,262 tons from 120 exporting nations over this period (Green and Shirley, 1999). The trade in dead coral peaked in the early 1990s but has since declined to levels comparable with the mid-1980s. The decline in trade is due to a number of factors including an export ban on dead coral from the Philippines, the primary source of coral in the 1970s. However, the quantity of corals traded live for the marine aquaria market has increased tenfold from 1985 to constitute more than half of the global trade in 1997, between 600-700 tons, and is continuing to increase 20-30% each year.

The live food fish trade is centered in Hong Kong, which imports a variety of species from Southeast Asia and the South Pacific archipelagoes at a wholesale value of $500 million per year (Lau and Parry-Jones, 1999). Specimens imported to Hong Kong are also re-exported to China, but little or none of this trade currently affects the U.S. (Lau and Parry-Jones, 1999). The U.S., however, does import significant amounts of traditional food fish (e.g., shrimp, spiny lobster, and queen conch) from coral reef ecosystems and thus has a responsibility to participate in their conservation as well.

The U.S. is a significant part of the curio and marine ornamentals industry: we import more than 80% of the stony coral in trade, and more than 50% of the marine aquarium fish (Anonymous, 2000). This trade is growing by 10-20% per year (Green and Shirley, 1999) and could have a substantial impact on reef health in the coming decade. In addition to a lack of science-based sustainable management, the marine ornamentals industry continues to experience destructive fishing practices (e.g., sodium cyanide use), harvest of multiple age classes within species, and poor survivability of traded specimens. The magnitude and growth of the U.S. market in the marine ornamental industry over the last decade all indicate that the U.S. should take a leading role in promoting sustainable use of coral reefs within range countries. As discussed below, several U.S. initiatives and international agreements with U.S. involvement are working toward this end.

Means to Address International Trade

The problems identified above are addressed in a number of arenas, including capacity building, cooperative management and planning, international regulation, and domestic U.S. measures. Each of these approaches addresses a slightly different aspect of international coral reef trade, and each has benefited from U.S. participation. This paper will note all of these areas, but will focus on the regulation of international trade, in both a CITES and non-CITES context.

Capacity Building

The U.S. remains committed to building the capacity for marine resource conservation in developing nations, particularly those that are coral reef range countries. Tropical marine ecosystems are crucial to our development plans in food security, the environment, economic growth, health, disaster mitigation, biodiversity conservation, and climate change (Anonymous, 2000). Many U.S. government agencies have committed funds and personnel to this effort, including the Department of State, the U.S. Agency for International Development, U.S. Fish and Wildlife Service, National Marine Fisheries Service, and the National Oceanic and Atmospheric Administration (Anonymous, 2000). Agency programs often focus on training or personnel loans to promote and implement coastal zone management, marine park administration, law enforcement, institutional decentralization, and fishery retraining. In addition, funds may be provided for partnerships with governments, non-governmental organizations, and academia.

International Cooperative Management

Since the mid-1990s, the U.S. has taken an increasingly active role in international fora aimed at coral reef conservation. These include the International Coral Reef Initiative (ICRI, an intergovernmental negotiation and planning forum), the Convention on Biological Diversity (which the U.S. has signed but not yet ratified; it includes active discussions over destructive fishing practices on coral reefs), the International Queen Conch Initiative, North American Wildlife Enforcement Group, World Customs Organization, ICPO-Interpol, Asian-Pacific Economic Cooperation, and the South Pacific Regional Environment Program. Most of these fora have focused on means to identify and reduce destructive fishing practices in tropical systems (e.g., sodium cyanide fishing), and the U.S. has participated actively through the commitment of funds, personnel, and expertise.

International Trade Regulation–CITES

The Convention on International Trade in Endangered Species of Wild Fauna and Flora (CITES) remains the primary instrument for monitoring and regulating trade in wildlife, including coral reef organisms. CITES is the only international legal mechanism with the mandate to protect species from overexploitation due to international trade. The opening paragraph of the Convention explicitly recognizes that each species of wild fauna and flora is an integral component of the natural ecosystem in which it is found, and it acknowledges that wild fauna and flora are to be conserved in perpetuity, in the wild. The ecological principles underlying this statement are reflected in Article IV of the treaty, which provides for its scientific underpinning.

There are now 154 countries that are Parties to CITES, making it one of the most extensive international agreements for species conservation and protection. CITES currently affects international trade in over 30,000 species of plants and animals. Over 2000 species of hard coral (all Scleractinia, organ pipe, fire, and blue coral), ten species of giant clams (Tridacnidae), and one conch species (Strombus gigas) are included in Appendix II of CITES, in addition to several non-reef building coral species. However, no species of coral reef fish are listed under CITES.

CITES entered into force more than 25 years ago, and benefits global species conservation through required scientific and management findings and action, annual species-specific trade statistics, international oversight on range country resource management, and a system of dual control in exporting and importing countries.

CITES offers three different levels of protection and regulation for listed species. Appendix I includes species threatened with extinction, which are or may be affected by international trade; commercial trade in these species is prohibited. Appendix II includes species that may become threatened if their trade is not brought under control. Commercial trade in Appendix II species is subject to regulation, and is allowed only if permits are obtained stating that trade will not be detrimental to the species. Appendix III includes species that are listed unilaterally by range countries, which require international cooperation in monitoring trade; commercial trade is permitted. All coral reef species covered by CITES are currently in Appendix II, which places obligations on both exporting and importing Parties. The treaty requires that each signatory nation establish a CITES Management Authority (to issue permits and perform certain other duties) and a CITES Scientific Authority (to monitor biological sustainability of trade).

CITES Article IV

Article IV of CITES provides the supporting framework and flexibility to scientifically evaluate the management of listed species 'Äì the very core of CITES. The export of Appendix II species requires a management finding that specimens to be exported were all obtained legally; it also requires a scientific finding that the export of the specimen(s) is not detrimental to the survival of that species. Once trade is underway, the status of the species must be monitored to ensure that it is maintained throughout its range at a level consistent with its role in the ecosystem.

All exports of specimens of CITES Appendix II species must comply with Article IV, which requires a two-tiered analysis for determining whether trade is biologically and ecologically sustainable. The first tier, Article IV paragraph 2(a), requires that the Scientific Authority determine that an export will not be detrimental to the survival of that species. The second tier, Article IV paragraph 3, further provides that the Scientific Authority must monitor the status of the species to ensure that it is maintained throughout its range at a level consistent with its role in the ecosystem and well above the level at which the species might become eligible for inclusion in Appendix I (which prohibits all trade). Conservation of species and their functional roles within the ecosystem would benefit from integrally linking Article IV paragraph 2(a) and a thorough implementation of Article IV paragraph 3 by the Parties; this is sorely lacking in most all cases.

Since Appendix II coral species constitute and create critical reef habitat, their abundance and health affects the overall vitality of the reef ecosystem. The linkage between Article IV paragraph 2(a) and paragraph 3 provides a check and balance mechanism to ensure sustainability of the Appendix II species in trade, and thereby the sustainability of the coral reef ecosystem. As the Scientific Authority fulfills its obligations under Article IV paragraph 3, non-detriment findings under Article IV paragraph 2(a) should take into account this scientifically based assessment of the species' role within its ecosystem and of the management of natural resources within the entire harvest site. Of course, it is a significant scientific challenge to effectively implement Article IV for coral species. There are uncertainties with regard to the biological and ecological sustainability of exploitation of coral reef ecosystem species, and controversy over the taxonomy of stony corals in the trade. Nevertheless, the listing of reef-building corals on CITES Appendix II was a decision of the CITES Conference of the Parties based on much conservation foresight, and provides unique opportunities and challenges to fully implement the scientific requirements of the treaty, by necessitating concomitant consideration of Article IV paragraphs 2 and 3.

Although these represent relatively simple concepts, such mandates require exporting countries to first develop and then examine, revise, or overhaul national management programs for listed species. Without robust and reasonable management schemes in place, range countries find it difficult to explain their permitted export levels, national licensing systems, or harvest quotas to other bodies within CITES.

Importing countries must ensure that each shipment is accompanied by a valid CITES export permit or refuse its acceptance. The U.S. Fish and Wildlife Service clears all wildlife shipments, and requires declaration from importers for all shipments containing wildlife, including coral and coral reef species. CITES permits are required for CITES-listed species, and a database is maintained for all of these imports. The U.S. reports these data annually to the CITES Secretariat, as part of its CITES obligations.

In addition, the treaty specifically authorizes CITES Parties to institute so-called stricter domestic measures, which may restrict otherwise permitted trade. These can include establishment of measures by importing countries to ensure that their imports from range countries are sustainable. For example, the European Union temporarily banned the import of six genera of stony corals from Indonesia after examining the high export quotas established by the Indonesian government. Indeed, the European Union has far-reaching legislation that requires import permits for all imports of CITES-listed species, and they can restrict imports by species or country, if they have information that CITES is not being effectively complied with by the exporting country. In particular, their scientific experts meet regularly to evaluate whether or not the Article IV sustainable use/non-detriment findings are being effectively made, and if not they can (and often do) stop imports of species from specified countries.

Such activity shows the impact of an Appendix II listing, and the powerful framework provided by CITES for monitoring and controlling international trade in coral reef species. The U.S. also has several stricter domestic measures that restrict imports of certain species based on conservation factors, including several CITES-listed species; these include the Endangered Species Act, Marine Mammal Protection Act, and Wild Bird Conservation Act. However, current U.S. law does not allow it to restrict imports of CITES Appendix II coral shipments, even if there are good reasons to believe that trade levels and practices are indeed in contravention of CITES' requirements, and are harmful to the species and its reef ecosystem. We will return to this concern in more detail later.

It is important to note that while the U.S. is the major importer of stony corals, commercial harvest of these species is banned or severely restricted in all U.S. range states and territories (Anonymous, 2000). This prohibition undoubtedly has many rationales, but U.S. policy makers often cite reef vulnerability as a prime factor in their decision. For example, the Caribbean Fishery Management Council (which develops management plans for several coral reef species in U.S. federal waters) has stated:

The principal value of coral reefs (including live-rock) is considered to be non-consumptive and are viewed as essentially non-renewable resource[s]...The importance of corals and associated plants and invertebrates lies in their relationship to the marine ecosystem. (CFMC, 1999)

Given the dichotomy between U.S. imports and U.S. domestic policies regarding reef management and our own trade, it is imperative that the U.S. ensures that it is not contributing to coral reef destruction in other parts of the world. The U.S. Coral Reef Task Force (USCRTF) accepted this challenge and requested that the Council on Environmental Quality (CEQ) lead an interagency team to investigate possible domestic measures (including trade restrictions) to address unsustainable U.S. imports of coral reef species (Anonymous, 2000). It is our personal view that it is important that our importation policies and practices are consistent with our domestic policies and management practices (regarding exports in particular). A case in point is the Wild Bird Conservation Act of 1992. When Congress was considering whether or not to restrict imports of CITES-listed wild birds, primary consideration was given to the fact that all commercial exports of native wild birds are prohibited under U.S. law.

U.S. Industry Efforts

The U.S. marine ornamentals industry has recognized the need to promote a sustainable supply of organisms harvested in a non-destructive manner from source countries. Toward that end, the Marine Aquarium Council (MAC) has been established as a non-governmental organization (NGO) to start dialog between collectors, wholesalers, retailers, and hobbyists on these subjects and to develop a means to certify environmentally sound products in the worldwide marketplace (Holthus, 1999). MAC intends to define, establish, and promote a certification scheme to be used throughout the chain of custody in the marine ornamentals trade, using best practice standards developed from multilateral consultations with industries, governments, and other NGO's (Holthus, 1999). The standards developed within MAC could help alleviate the poor shipment survival of many marine ornamentals, and may attract more consumers to "environmentally friendly" products.

Obstacles to overcome in this process include market economics (uncertified, cheaply collected specimens will cost less than certified ones) and industry consensus. Obviously, as an industry organization MAC's interest is to avoid additional regulatory action or any legislative restrictions on U.S. imports. However, MAC has been actively involved in the U.S. Coral Reef Task Force discussions for many years, has been a positive partner, and has provided useful advice to U.S. agencies addressing these issues.

Domestic U.S. Measures

The USCRTF International Working Group appointed a small Trade Subgroup to examine the CEQ process and develop possible measures to channel U.S. demand toward sustainably harvested reef species. The subgroup recommended a variety of actions for the U.S. to consider, including:

1. The U.S. should prohibit domestic harvest or collection and the import or export of coral reef species and products listed under CITES that are not sustainably managed or from environmentally sound mariculture programs. Exceptions might include organisms intended for approved captive breeding programs, scientific research, or public display. Restrictions might be extended to other species of concern under certain circumstances.
2. The U.S. should work with members of the marine aquarium industry, environmental organizations and other stakeholders to develop, within a specified time limit, responsible practices and guidelines for collection and transport of coral reef species that reduce mortality rates throughout the trade stream, improve product quality, and ensure survival in captivity.

3. After working with stakeholders over a specified time period, the U.S. should prohibit the domestic harvest and collection of any coral reef species by defined destructive fishing practices, such as the use of reef dredging, explosives, or poisons.
4. After working with stakeholders over a specified time period, the U.S. should prohibit the import or export of any coral reef species unless accompanied by certification that the products were not taken through the use of destructive fishing practices.
5. The U.S. should develop a coordinated national strategy for conservation and sustainable management of coral reef species and ecosystems within the U.S., and then work with the international community to share this strategy and develop criteria for the conservation and sustainable management of coral reef ecosystems in other parts of the world (Anonymous, 2000).

All of these are important options to consider. Such efforts would be intended to channel the U.S. demand toward source countries with effective management plans and/or those species that are harvested sustainably.

The legislative option would create a legislative and regulatory situation comparable to the U.S. Wild Bird Conservation Act, which supports CITES by allowing importation into the U.S. of certain approved captive-bred species, but also allows for importation of wild-caught birds from approved science-based sustainable use management plans. It is a "reverse list" approach. The WBCA, passed in 1992, was stimulated by unsustainable trade, and challenges to effective CITES implementation by range countries, similar to what we are seeing now with the trade in CITES Appendix II coral species. The WBCA also encouraged the rapid development of captive breeding and husbandry techniques analogous to those that are emerging in the marine ornamentals industry today. Such a legislative option for U.S. imports would certainly benefit from more discussion between Congress, the Administration, scientists, conservation organizations, industry, and others.

Options that relate to cooperative work with industry are already underway, working through MAC and others. Certainly best management and trade practices are vital to both ensuring sustainable trade, and to conservation of coral reef ecosystems. We personally believe that such "voluntary" efforts are often most successful when coupled with regulatory/legislative solutions and incentives, and as such further dialogue on these options would be very useful.

Finally, all of these options address U.S. import (and export) policy options, and options available to the export and import industry. We believe that it is vital to couple those strategies with increasing public awareness. We strongly urge efforts to create informed consumers, so that anyone purchasing coral jewelry, coral reef-based curios, coral for a home aquarium, or ornamental fish, should be able to be aware of the origin of the products they are purchasing.

We are not recommending consumer boycotts of coral reef-based products. Rather, consumers should be educated to demand products that are produced in a biologically, culturally, and economically sustainable manner, and provide both economic benefits to local people, and incentives for coral reef ecosystem conservation. An informed consumer is powerful--and can guide industry best management practices, government policies, and of course Congressional interest and action.

OCEAN ATTITUDES 2001: CONSERVATION THROUGH CONSUMER ACTION

Setting the Context

To assess public attitudes regarding a variety of ocean issues, between 1996 and 2001, SeaWeb has worked with professional market research firms. The results reported here represent a subset of the opinions expressed by survey respondents. In all cases, the surveys were conducted using industry-standard market research techniques and included sample sizes that allowed for accurate calculation of statistical significance.

Most Americans care about the oceans and believe that the health of oceans is threatened. Fifty-nine per cent of those responding to a 1999 poll rated the overall health and quality of the oceans as "only fair" or "poor," while less than one-third thought the oceans are in "good" or "excellent" condition. The same poll showed that over half believe that the condition of the oceans is deteriorating, having gotten worse in recent years. An even larger percentage 'Äì nearly three-quarters 'Äì evaluate the condition of coastal waters as negative. Coastlines are where the majority of people interact most directly with oceans.

Studies consistently show that the public--caring most about human health--views pollution with greatest concern and as the topmost threat to ocean health.

- From among a range of environmental problems--water pollution, rain forest loss, development, overconsumption, etc.--people consider most urgent those with the most direct impact on human health, namely water pollution and toxic wastes.
- The 1999 poll gave people a list of specific ocean issues--ranging from oil spills, corporate farm runoff, and improperly treated sewage to coastal

development, commercial overfishing, and damage from boating/jet skis--and asked respondents to rank them. Deterioration of coral reefs was a middle tier concern, ranked well below pollution concerns but above such concerns as damage from boating and recreational overfishing. While this places coral reefs in a context of other concerns, it is difficult to know whether people would have themselves volunteered reef loss in the survey.

- In 1996, while testing messages that would help communicate more effectively about oceans, SeaWeb found that people connect to the need for ocean conservation mostly through two themes--a sense of responsibility to future generations and concern for human health. Making the case for coral reef conservation could use examples that fit these themes.

Communicating about specific threats to coral reefs and solutions is complicated by public confusion about reef systems. Approximately four in ten Americans either believe that fish breeding grounds and coral reefs are found throughout the oceans (versus only in certain places, as is the case) or do not know.

Towards Coral Reef Conservation and Consumer Action

However, there is strong support for establishing marine reserves. Respondents in the 1999 poll included protection of ocean life and habitats--and coral reefs specifically--as some of the top goals for marine protected areas. These were placed well above such other goals as responsible management for oil exploration/drilling and commercial fishing, maintaining clean areas for swimming and diving, and providing recreational areas for boating. Nearly three-quarters of the public would support prohibiting the collection of tropical fish and corals from protected areas.

The public clearly is concerned about the state of the world's oceans. The public believes that humans can do lasting damage to the oceans, and supports efforts to strengthen ocean protection. However, many do not believe that their personal actions have much impact on ocean health. On the other hand, taking environmental action as a consumer--for example, recycling motor oil, cleaning up litter, eating only environmentally safe fish, and buying non-overfished seafood--is more popular than other kinds of individual action (such as contacting politicians and joining an environmental group).

Indeed, there are substantial numbers of people willing to modify their purchasing behaviors to help oceans. For example, many strongly support actions to protect oceans, even if it meant paying more for seafood. Sixty-two per cent of respondents in one survey said they would not eat fish classified by the government as overfished, and 44 per cent would only eat fish caught or farmed in a way that protects oceans.

Americans are increasingly connecting to conservation through consumerism and there is growing momentum to use consumer markets to drive ocean conservation.

- The Marine Stewardship Council has developed a certification program for sustainable fisheries.
- Various organizations, like the Monterey Bay Aquarium, National Audubon Society, and the Chefs Collaborative have developed seafood-buying guides for consumers, chefs, and restaurateurs.
- SeaWeb has started the Seafood Choices Alliance as a trade association to provide the seafood sector with the information it needs to make sound choices and provide the best options for its customers.

For coral reefs, there are two areas of consumer concern--food fish and marine ornamental fish, coral and live rock. To address the consumption of fish, the Seafood Choices Alliance and other organizations are working to create a sustainable food fish industry. To address the trade in ornamentals, the Marine Aquarium Council seeks to create a sustainable trade through certification of marine ornamental fish.

In addition to formal opinion polling, SeaWeb conducted an informal survey of 77 aquarium hobbyists attending a conference of the Marine Aquarium Societies of America. The results of this survey suggest that there is great interest in using purchasing behavior to support reef stewardship. Most respondents indicated they want to support an industry based on quality and sustainability. Some hobbyists currently seek out suppliers of healthy animals, and are "very interested" in the source and collection methods of the fish they purchase. More importantly, they are willing to pay more for fish that are certified as being caught and handled in an environmentally responsible way. While based on a small, self-selected sample of hobbyists, these results point to a positive atmosphere for providing information that would allow marine aquarists to make environmentally sound choices.

Conclusion

There is a need to better understand the public's knowledge about and attitudes towards coral reefs so that campaigns and communications can be crafted more effectively. There is the opportunity to use consumer initiatives as a tool to augment legislative and enforcement solutions to the coral reef crisis. This will require communicating the link between reef conservation and informed consumer purchases--a difficult task in the face of public confusion about coral reef ecosystems and how they work.

However, saltwater aquarium hobbyists and the trade can be a force for change. And, with growing interest in linking consumer behaviors to conservation, there needs to be continued investigation of how environmentally responsible aquaculture--whether for food fish or ornamentals--can relieve pressure on coral reefs without unintended consequences for reef-based economies.

INTRODUCTION TO THE SUSTAINABLE DEVELOPMENT CONCEPT IN FISHERIES

Background

In fisheries, development may be defined as "a process of change through which sustainable and equitable improvements are made to the quality of life for all or most members of a society" (Bailey and Jentoft, 1990). Adding the need for sustainability implies that these improvements need to be achieved without risk to the long-term stability of the ecosystem concerned.

In 1987, the World Commission on Environment and Development defined sustainable development simply as "Development that meets the needs of the present without compromising the ability of future generations to meet their own needs." This definition recognises that human needs can change with time, and that preservation of the integrity of the environment is necessary to ensure that future generations can realise the full potential of their own development needs. Another, more elaborate definition states that "Sustainable development is the management and conservation of the natural resource base, and the orientation of technological and institutional change in such a manner to ensure the attainment and continued satisfaction of human needs for present and future generations. Such development conserves land, water, plant and genetic resources, is environmentally non-degrading, technologically appropriate, economically viable and socially acceptable." This definition was developed by the FAO Committee on Fisheries in 1991, and acknowledges the need to achieve a workable balance between guaranteeing satisfaction of both present and future human needs including fulfilment of social and economic demands, and conserving the natural resource base. In response to the general goals identified, FAO developed a Code of Conduct for Responsible Fisheries (FAO, 1995) that provides principles for ensuring sustainable exploitation of marine resources.

Searching for Optimum Fisheries Management

The 1991 FAO definition noted earlier recognises the need for management of the resources concerned as one of the essential actions for achieving sustainable development. To accommodate the other components of the FAO definition, it is clear that such management must embrace a range of objectives, incorporating biological, economic, and social elements. Knowledge of the status of the resources and the likely impact of fishing activities and other factors on these resources provide for informed and successful management of fisheries.

In the early days of fisheries management, scientists focused on the biological modelling of fish stocks (Graham, 1935; Beverton and Holt, 1957). Assuming that fish stocks respond predictably to moderate levels of exploitation, with a defined equilibrium state, fisheries scientists are able to calculate the maximum level of catch that would allow sustainable exploitation. This sustainable maximum catch is referred to as the Maximum Sustainable Yield (MSY), and continues to be a key reference point in present-day fisheries management (Gulland, 1983).

MSY takes into account only the sustainability of the resource and so satisfies only a biological objective. In view of this, fisheries economists have argued that fishing is conducted as a business and that the economic benefits of the activity must not be ignored. This led to the development of another management reference point, referred to as Maximum Economic Yield (MEY). MEY is calculated based on optimising the difference between the cost of fishing and the income gained, and usually occurs at a lower fishing effort than the fishing effort occurring at MSY (Hersoug, 1996).

Subsequent to the development of MEY, social scientists have noted the importance of addressing social concerns, to ensure more equitable distribution of economic benefits. The social objective of Maximum Social Yield (MsocY) was then recognised, and the associated reference point varies with the fishery situation.

These biological, economic and social management objectives, as currently defined, are now facing yet another challenge, with the development of chaos theory (Smith, 1990 in Symes, 1996; Wilson and Kleban, 1992 in Symes, 1996). Chaos theory suggests that although nature is non-random, it is unpredictable. The equilibrium state of fish populations assumed by MSY is not accepted, and fish populations are believed to vary unpredictably within limits (Symes, 1996). Chaos theory therefore proposes 'ecologically adapted management', based on knowledge of the long-term stable ecological relationships, and requires more flexible management systems.

Achievement of Responsible Fishing

Sustainable fisheries development can be achieved through responsible fishing, which considers rational fishery management objectives that address a range of issues including the status of the resource, the health of the environment, post-harvest technology and trade, as well as other economic concerns, social benefits, legal and administrative support. In the case of shared resources, a co-ordinated approach to responsible fisheries management is essential, and Caddy and Griffiths (1995) proposed the following actions:

Regulate Fishing Effort

It is crucial to control fishing effort, and to avoid financial incentives that would contribute to excess fishing capacity. Excess fishing capacity and overcapitalization threaten the sustainability of the resource, as well as the industry.

Establish Code of Conduct for Responsible Fishing to Guide Management Plan

This is needed to maximize benefits from the fishery, while avoiding wastage caused by indiscriminate fishing practices. In particular, it is important to reduce catches of undersized fish and non-target species, and to avoid use of gears that have a negative impact on the environment.

Establish and support regional/international fishery commissions and organisations concerned with management of shared resources

The successful management of shared resources requires effective coordination by all the countries concerned. The relevant international fisheries agreements promote participation in, and financial support for, the work of these commissions and organisations.

Regular Consultation among Harvesting Countries

Parties sharing the resources need to consult and collaborate regularly so as to promote understanding and full cooperation. Set agreed management objectives and related reference points, incorporating a precautionary approach

Agreement on management reference points during the early stages of the fishery will help to ensure full cooperation of participants with management decisions. Where there is scientific uncertainty, a precautionary approach to management is recommended.

Develop Contingency Plans

Management plans should incorporate some contingency for dealing with sudden and unpredictable environmental changes caused by man-made or natural disasters.

Develop Mechanism for Resolving user Conflicts

Management should provide mechanisms for handling problems arising from resource user conflicts.

Protect Biodiversity

The biodiversity or species richness of an ecosystem is an important measure of ecosystem health. The preservation of biodiversity will ensure that present human development activities do not threaten the ability of future generations to meet their own needs.

Protect the Environment

There should be monitoring and control of waste disposal and pollution. In addition, every effort should be made to prevent discarding of entangling material that could trap and kill species or physically damage the environment.

Promotion of Research

Research should be conducted to support and inform various aspects of management.

Optimise Social and Economic Stability

There should be fair and equitable distribution of benefits derived from the fishery.

Constraints to Caribbean Island States

For many small developing island states within the Caribbean, there are a number of constraints that pose significant challenges to the sustainable development of fisheries, including the following:

- *Fishing Practices:* Destructive fishing methods such as dynamiting fish are still practiced in certain areas. Additionally, fishers continue to use illegal mesh sizes that catch very small fish, and non-target species that may have little or zero market value.
- Inadequate knowledge of the resource and ecosystem

  Many Caribbean islands do not have sufficient detailed data to permit an accurate evaluation of the status of their resources and a good understanding of their marine ecosystem.
- Inadequate knowledge and recognition of social and economic conditions

  Many governments within the region still do not have an adequate appreciation of the social and economic potential of sustainable fisheries development, and hence still invest minimum resources in fisheries development and management activities.
- *Absence of Long-term Policies:* The lack of long-term policies makes it difficult for Caribbean countries to maintain pace with rapidly evolving trends in global fisheries management approaches and trade strategies.

- *Capacity:* In many Caribbean countries, fisheries administrations are poorly staffed, and there are limited numbers of skilled and knowledgeable fisheries technicians, scientists and managers, and limited equipment and funds available for basic tasks such as data collection and research. There are few research institutions and regional organisations within the region, and some of these suffer similar limitations in available funding and a broad range of technical expertise.
- *Resource user Input:* In the Caribbean, fisherfolk organisations have not actively participated in the management process, mainly owing to a lack of good organisational and administrative skills. On the part of the fishers, this has resulted in a lack of trust in, and respect for, the governments that are responsible for making fisheries management decisions.
- Capability for monitoring, control and surveillance

  In many instances, there is little or no capacity for monitoring, control and surveillance activities, and no associated legislative framework.
- *Post Harvest Aspects:* Caribbean countries are just beginning to develop their export markets for fish and fish products, and face the challenge of satisfying stringent standards in all aspects of fish handling, processing and packaging, recently established by the importing countries.
- *Facilities:* Equipment and facilities are not upgraded regularly to deal with expanding fishing activities and increasing management demands.
- *Habitat Degradation and Pollution:* For some time, there have been signs that the Caribbean ecosystem is under stress (Richards, and Bohnsack, 1990).

Degradation of habitat, particularly in the coastal areas has resulted from uncontrolled coastal development activities mostly associated with expanding tourist industries and overpopulation problems, as well as extensive sand mining and deforestation.

- *Financial Resources:* Many Caribbean countries are small island developing States with limited financial resources. This impacts negatively on their ability to cope with the wide range of issues required for successful sustainable fisheries development.

Conclusions

Caribbean governments need to appreciate the social, economic and financial potential of sustainable fisheries development. Fisheries administrations need to be given more financial resources, and properly staffed and equipped to address effectively and completely all aspects of sustainable fisheries development and management. Additionally, small island states should recognize the benefit of, and work towards, regional coordination in fisheries management activities, including sharing of expertise and resources for education, research, technology, monitoring, control and surveillance activities, and development of the relevant legislative framework.

Having removed constraints to responsible fishing, fisheries management must then strive to achieve a workable balance between defined objectives protecting the resource and its environment, and those seeking fair and equitable distribution of viable economic benefits.

CURRENT PROBLEMS IN THE MANAGEMENT OF MARINE FISHERIES

The public perception of fisheries is that they are in crisis and have been for some time. Numerous scientific and popular articles have pointed to the failures of fisheries management that have caused this crisis. These are widely accepted to be overcapacity in fishing fleets, a failure to take the ecosystem effects of fishing into account, and a failure to enforce unpalatable but necessary reductions in fishing effort on fishing fleets and communities. However, the claims of some analysts that there is an inevitable decline in the status of fisheries is, we believe, incorrect. There have been successes in fisheries management, and we argue that the tools for appropriate management exist. Unfortunately, they have not been implemented widely. Our analysis suggests that management authorities need to develop legally enforceable and tested harvest strategies, coupled with appropriate rights-based incentives to the fishing community, for the future of fisheries to be better than their past.

The United Nations Food and Agriculture Organization (FAO), which monitors the state of world fisheries, has estimated that since 1990 approximately one-quarter of fish stocks have been overexploited, depleted, or are recovering from depletion (17%, 7%, and 1%, respectively), with the Northeast and Northwest Atlantic, the Mediterranean, and the Black Sea being the areas with the largest number of depleted stocks. Many authors have elaborated on these conclusions, documenting the poor state of fisheries worldwide. Nevertheless, the situation, although serious, is not catastrophic, and there are grounds for optimism. There have been successes of fisheries management, and there is an understanding of what is involved in successful fisheries management and of the requirements for its implementation. These issues are explored in this review.

The management of commercial fisheries clearly requires a good scientific understanding of the behavior of the exploited stock or stocks. The science that is used to assess commercially exploited species is still dominated by the population models developed by Beverton and Holt for single-species assessments some 50 years ago. The availability of substantial computing power has meant that sophisticated estimation methods can be used, and an appreciation of the way in which fish stocks respond to environmental variability is readily incorporated in scientific advice. Calls for a more ecosystem-orientated approach have been voiced for some while, but the paucity of data and the demands of multiparameterized multispecies models means that most ecosystem considerations in practical stock assessment tend to be ad hoc manipulations of the single-species approach.

What has developed is a realization that effective management requires an understanding of how the fishery system is performing relative to reference points. The most commonly used reference points are those relating to the size of the stock itself and the fishing mortality that will result in these stock sizes, given existing relationships between the stock, recruitment, natural mortality, and growth (Fig. 1). A typical "target reference point" is the biomass necessary to produce maximum sustainable yield (BMSY). However, such targets do not explicitly recognize threats to the stock. To address this issue, stock size "limit reference points" are usually defined or interpreted as the stock biomass below which recruitment becomes substantially reduced. Clearly, it is important to avoid situations where the stock is at or below this level. Accordingly, management should aim to have as a target a level of stock size that carries a low risk (allowing for scientific uncertainty) of the stock dropping below the limit reference point. This could mean having a target level of fishing mortality that provides stock sizes above BMSY.

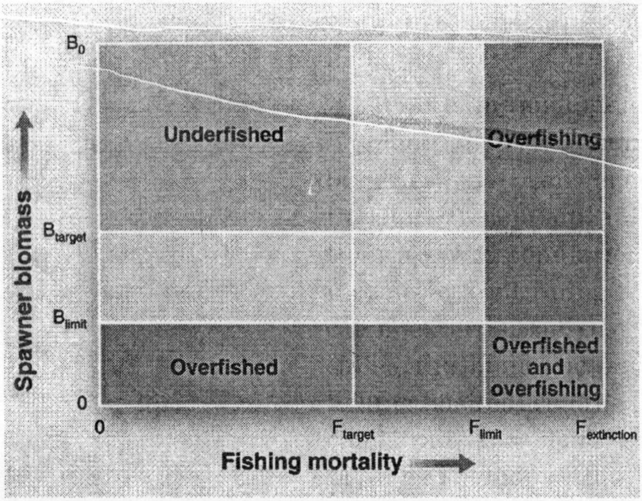

Fig. 1. Typical reference points and stock status definitions for stock biomass and fishing mortality. The limit of fishing mortality that generates biological extinction is F extinction.

Understanding Fisheries Management

Competent scientific advice based on appropriate data is far from ubiquitous in the fisheries world, and even in ideal situations, fisheries management has often been unsuccessful. The success of a management system is often defined in terms of biological, economic, social, and political objectives. Clearly, economic and social objectives will not be met while a stock is in such a depleted state that the long-term sustainability of the fishery is threatened, but equally, biological objectives are unlikely to be met without consideration being given to economic and social objectives.

Hence, we argue that an understanding of the fishery management process can only come from analyzing the capacity and incentives of the two key stakeholders: the fishing community and the management authority. This is not to belittle the importance of other stakeholders, such as recreational fishers and environmental groups, who have important roles in the management of certain fisheries.

Where management is weak or nonexistent, the economic factors underlying overfishing in commercial fisheries have been generally understood since the 1950s. In short, when multiple fishers compete to catch fish from a given population, each fisher maximizes his net income by continuing to fish as long as the value of his catch exceeds the cost of catching it. An equilibrium, called the bionomic equilibrium, is reached only when fishing has reduced the fish population to a level at which catch rates are barely sufficient to cover the costs of fishing. The population is then maintained at this level through biological processes of natural growth and reproduction. Thus, if the price:cost ratio is high, the bionomic equilibrium will result in a low stock of fish, and hence a low annual catch level; two characteristic features of overfishing. In addition, the so-called economic rents (total revenue minus total costs) from the fishery will equilibrate at zero, resulting in minimal overall economic efficiency.

Many management authorities seek to meet their objectives by setting output controls in terms of a total allowable catch (TAC) for the year and closing the fishery when the year's cumulative catch has reached the TAC. Restrictions on fishing gear, fishing season, and fishing areas, as a supplement to the TAC, may also be imposed. If the TAC is correctly specified and enforced, this method should maintain a stock level well above that of bionomic equilibrium. If, however, the TAC and the science behind it are not respected by fishermen and not adequately enforced by authorities, widespread illegal fishing can occur. A recent example is in the eastern Baltic cod fishery where illegal fishing contributes to true catches being some 35 to 40% higher than reported.

Efficient enforcement can be difficult. Put simply, fishers will be deterred from breaking fishing regulations if their expected loss from detection and successful prosecution exceed their expected gain. In many fisheries, the probability of detection of illegal activity and the penalties are not sufficiently high to act as a disincentive.

Strong management can ensure that biological targets are met, but it is essential that regulations are enforceable, and this has often proved to be difficult. Less-than-perfect enforcement can lead to illegal fishing, poor scientific data, and a failure to meet biological targets. Input measures, such as limiting the number of vessels or restricting available season length, are usually more easily enforceable than output measures such as TAC. However, control via input measures is vulnerable to effort creep, whereby operators increase the fishing power of their vessels through technical means. Nevertheless, monitoring of vessel performance over time and adjusting the allowable level of effort have allowed successful effort control to be implemented.

Overcapacity

Simplistically, it would seem that positive economic rents should also emerge in a TAC-regulated fishery. In reality, many TAC-regulated fisheries have experienced an unexpected increase in fishing capacity, as additional vessels enter the fishery in response to (temporarily) positive rents. Economic models then predict a regulated bionomic equilibrium, in which economic rents (net of fixed costs) again equilibrate at, or near, zero. This situation currently exists in many of the world's regulated fisheries; overcapacity of fishing fleets is widely perceived as a major impediment to achieving economically productive fisheries. It is thus ironic that such overcapacity is usually generated by the management system itself, although it can also result from high profitability during the initial phase of a newly developing fishery.

Overcapacity is widely recognized as a major problem affecting world fisheries. With its attendant social and economic problems, overcapacity can, via the political process, lead to the erosion of management control. It is also understood to be one of the results of subsidizing fisheries, which even today is estimated to be several tens of billion U.S. dollars per year. Such subsidies directly undermine the sustainability of fisheries because they lead to a bioeconomic equilibrium with high levels of fishing and low stock size. In several fisheries, government funds have been used to buy out excess fishing capacity. For various reasons, such buyback programs have been less effective than expected. First, often only the least efficient vessels are bought up, leaving total fishing capacity largely intact. Second, the buyback program by itself does not remove the economic incentives underlying overcapacity, which tends to increase once the buybacks are completed.

Thus, the underlying cause of the dual crisis of overfishing and overcapacity, as well as other undesirable outcomes, such as habitat destruction and incidental kills of untargeted species, can be found in the economic incentives of fishers who compete for their annual catches. These incentives are not affected by management strategies that retain the competition between fishers for a common-pool resource. Perhaps the most important development in fisheries management over the past 20 years has been the recognition of this fact and the introduction of rights-based management in several regimes. Indeed, it has been argued that of the tools at the disposal of managers, more emphasis needs to be placed on incentive-based approaches that better specify community and individual harvest or territorial rights, in additional to public research, monitoring, and effective administrative oversight.

Transferable Quotas

An alternative management strategy based on individually allocated transferable annual catch quotas (ITQs, or individual transferable quotas) is now in effect in several fishing nations, including Australia, New Zealand, Iceland, Canada, and Namibia. A well-organized rights system alters the economic incentives of fishers, who no longer compete for their catches, so that highly competitive fishing no longer takes place. The guarantee to fishers of a certain proportion of the catch allows them to make rational economic choices about where and when they catch fish. An ITQ system goes further, allowing the industry to settle on a fleet capacity that optimizes individual economic yield to vessels or cooperatives, although this of course can still be distorted by inappropriate subsidies.

In addition, ITQ fishers may often be expected to favor management actions that protect and enhance fish populations, because the value of a quota share increases as stocks become more abundant. Problems that may arise, such as misreporting or high-grading of catches, have been successfully countered by the use of observers, required by the management system but paid for by the industry; observers are used extensively in the U.S. Pacific fisheries, Australia, and New Zealand. Experience with ITQ systems shows that many fishers willingly support and adhere to conservative management strategies and may also avoid fishing practices that endanger habitat or threaten other species, so long as they are guaranteed long-term rights. But this does not mean that enforcement and scientific monitoring are unnecessary in ITQ systems; both are essential unless catch levels are set at precautionary low levels. It is thus unsurprising that the two countries with perhaps the most fully developed ITQ systems, New Zealand and Iceland, have some of the highest costs of management per fishing vessel.

Several authors have pointed to instances of successful fisheries management in both the developed and developing world. Among their conclusions are that incentive structure, institutional capacity, and participation of stakeholders are of key importance. However, in some studies, a rights-based approach is seen as the primary mechanism to deliver this, whereas in others, severe top-down controls with very limited participation of fishing communities in the management process are advocated. We argue here that a necessary condition for successful management contains all these elements: a competent management authority able to set and enforce regulations and monitor the status of the stock, together with some form of rights-based allocation to fishing operators (either collectively or individually) to avoid the situation where overcapacity produces economic hardship and erodes management capacity.

Evidence from Fisheries Performance

Reviews of successful fishery management are of necessity specific to individual fisheries and sometimes anecdotal. However, in some large areas, a combination of strong state governance and wealth, substantial scientific activity, and different types of fishery management offer the opportunity of some comparison between different types of fishery management that goes beyond the anecdotal.

Detailed data on the status of different fisheries are published for U.S., Northeast Atlantic, Australian, and New Zealand fisheries. The approach to management taken by these authorities is varied. New Zealand has the most developed and widespread application of individual user rights (ITQs), which have been in place from 1986 and have spawned other developments such as collaborative and alternative research by stakeholders. ITQs are present also in some Australian fisheries, a very few U.S. fisheries, and some Northeast Atlantic fisheries. (The Faroes, Norway, Iceland, and United Kingdom have rights-based systems, and some fleets, notably the Dutch flatfish and Spanish Grand Sole fleets, are also managed via ITQ.)

If a broad view is taken of these management areas, the evidence for the positive benefits of ITQs in supporting sustainable resource use is mixed. Only 15% of New Zealand's stocks within the quota management system, for which the stock status is known, are substantially below the target reference level. For other administrations, the percentage of stocks that are below the limit reference level (i.e., overfished), out of the total number of stocks for which the status is currently known, is 19% for Northeast Atlantic fisheries managed by non-European Union (EU) administrations (Iceland, Faroes, and Norway), 25% for federally managed U.S. fisheries, 30% for Northeast Atlantic fisheries managed primarily by the EU, and 40% for Australian Commonwealth fisheries. Even within the United States, there are very large regional differences: 40% of major Fish Stock Sustainability Index (FSSI) stocks managed by the New England and Mid-Atlantic Fishery Management Councils, for which the stock status is known, are overfished, and 30% are subject to overfishing. By contrast, only 13% of FSSI stocks managed by the Pacific, West Pacific, and North Pacific Management Councils are overfished, with 6% suffering from overfishing.

Only the United States has seen an improvement in performance over the past several years; in 2000, 38% of

U.S. stocks for which the status was known were classified as overfished. All other areas have experienced some increases in the number of overfished stocks in the past decade, although the increases in New Zealand have been very small and are offset by recoveries in some inshore stocks. However, these statistics disguise a quite dynamic situation within each region; for example, of 74 U.S. stocks requiring rebuilding, biomass is increasing in 48% of them even if they have not yet achieved rebuilt status.

More detailed examination of the U.S. situation reveals that although ITQs are not generally applied, West Coast fisheries are managed by quota controls with fishing rights assigned to fishing companies or sectors, whereas in the northeast, fisheries are managed by a days-at-sea scheme and other effort controls. In terms of their current performance and the stock recovery required by the Magnusson-Stevens Act, West Coast management systems appear to be more effective than northeast coast systems: Only two of the 18 New England stocks that were overfished in 1995 have now recovered, compared to 4 of 9 stocks similarly categorized by the Pacific Fisheries Management Council.

Clearly, non-ITQ management systems do not always fail to maintain sustainable stocks, and management systems using ITQs are not always successful. The critical additional requirement appears to be a formally adopted management strategy with predefined rules for what to do in different circumstances. In New Zealand, in addition to an ITQ system, a formal harvest strategy embedded in the Fisheries Act (1996) means that rebuilding is statutorily required when the stock is below its target level. By contrast, the lack of a formally adopted harvest strategy in the Australian Southern and Eastern Scalefish and Shark fishery has led to an increase in the number of overfished stocks in that fishery over the past 10 years, despite operating with an ITQ system since 1992. EU fisheries also lack a formal harvest strategy; although there is a commitment under the 2002 revision of the Common Fisheries Policy to develop multiannual management plans for all stocks, such plans are only currently defined for 17 of the 94 stocks that fall under EU management, and many of these have had to be negotiated during periods of stock collapse.

The key problems, i.e., the need to provide incentives to fishers to engage constructively in fisheries management and the need to have strong legal support for predefined harvest strategies, apply equally to management of stocks under national control and those in international waters under the control of Regional Fisheries Management Organisations. To our knowledge, none of the latter currently allocate rights to individual fishers, and only a few have defined and tested effective harvest strategies. Allocation problems continue to beset these high-seas fisheries and influence compliance, data availability, and transparency.

The recovery of depleted fish stocks is a key issue and one to which most countries committed themselves in 2002 as part of the World Summit on Sustainable Development. An effective reduction in fishing effort, the participation of fishers and other stakeholders in the science and decision-making process, and the biology of the species are important factors affecting successful recovery. However, unless a harvest strategy is defined, with pre-agreed, legally binding decision rules requiring reductions of effort when stock sizes decline below limit reference points, most management authorities will still delay taking action to recover stocks. Some of this delay may arise from uncertainty in the science, but mostly it arises from an unwillingness to take decisions that will create hardship for fishers, and usually a delay will exacerbate stock decline. Ultimately, the most successful management approaches are likely to combine rights-based systems, creating incentives for fishers to operate efficiently and with long-term sustainability in mind, with a strong legal structure that requires the development of pre-agreed harvest strategies and decision rules that are triggered and adhered to as reference points are passed. As indicated earlier, an adequate control of fishing activities is also necessary.

ADDRESSING ECOSYSTEM-BASED MANAGEMENT

In recent years, there have been many calls for much wider use of Marine Protected Areas to address the need for ecosystem-based management. We see these as a useful part of fishery regulation, but they are not a universal solution because unless the basic issues of capacity, regulation, and rights are solved, protected areas will simply displace the problem elsewhere. In this we concur with recent reviews that emphasize the primary importance of conventional measures to control fishing mortality and the secondary but essential role that marine protected areas (or other area-based management, such as local prohibitions on particular gears such as bottom trawls) have in dealing with specific issues of ecosystem conservation, such as bycatch and habitat damage.

The simple creation of rights-based incentives does not automatically deal with ecosystem problems, because fishers have little incentive to minimize bycatch or habitat damage that does not affect their target species. An interesting recent development is the creation of additional incentives for fishers through market measures, such as the creation of sustainable fisheries certification schemes and pressure from environmental nongovernmental organizations for responsible fisheries. Fishers have a major incentive to improve fisheries to satisfy certification conditions, and so far most of the conditions raised in Marine Stewardship Council certifications have concerned the ecosystem effects of fishing, often related to quantifying and reducing deaths of bycatch species and damage to habitat. Even in the statistics documented for some of those states with appreciable management capacity, what is striking is for how many stocks, the status is uncertain or not determined. In the United States, the stock status of 30% of the 230 major (FSSI) stocks and stock complexes was undetermined in 2006; in Australia (48%), New Zealand (78%), and the Northeast Atlantic (61%), the numbers are even higher.

Given the problems that most authorities have in deriving reliable quantitative assessments of their stocks of major commercial importance, the large numbers of small, commercially unimportant stocks present in most areas, usually as bycatch, cannot realistically be assessed. Under a comprehensive ecosystem approach, risk assessment methodologies should be used to identify those bycatch species in need of special measures, and monitoring programs, for instance using scientific observers, need to be implemented to monitor trends in all bycatch species. The application of these approaches is in its infancy even in the most advanced management schemes; many simply respond by setting untested but hopefully precautionary effort or catch limits.

These considerations apply even more strongly to fisheries operators in developing countries. In a situation of little or no management capacity, some form of bioeconomic equilibrium is the likely result, but in such cases the management priorities may be different. Indeed, high employment with relatively modest economic rent, as long as it is compatible with the sustainability of the resource, may be a perfectly legitimate management goal (Fig. 2). In other cases, the development of Territorial

Use Rights (TURFS) within local communities can lead to effective management control and rights-based operations, resulting in successful management (20, 21).

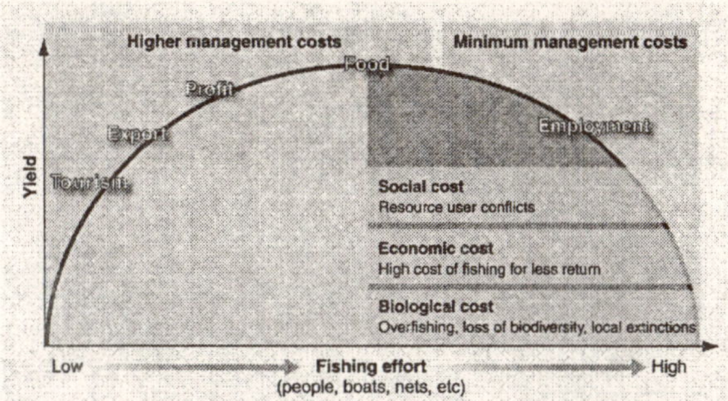

Fig. 2. The fishery management dilemma is illustrated with a simple stock production curve showing sustainable yield varying with effort. Low effort reduces biological risks and enhances economic profits at the cost of low employment and higher management costs. High effort increases employment at the cost of low economic profits and increased biological and social risks, but with low management costs.

Concluding Remarks

There is no doubt that there is a major problem with the world's fisheries, and, despite serious attempts to improve management and to facilitate recovery of depleted stocks, the success has been limited. The key issue that we highlight in this review is that for successful management a dual approach is required, one in which authorities provide incentives for conservation based on fishers' rights and which is supported by strong management incorporating legally enforced and tested harvest strategies.

COMMUNITY-BASED FISHERIES MANAGEMENT IN THE BAY OF FUNDY: SUSTAINING COMMUNITIES THROUGH RESISTANCE AND HOPE

Communities along the Atlantic Coast have relied on the fishery resource for centuries. Recent declines in fish stock, exclusion of First Nation communities, and government policies that favor high-cost allocation of the fishery have threatened the economy, vitality and well-being of coastal communities. This case study describes several powerful tools that were used to gain access to the fishery by local communities. These tools include organization, protest, and litigation. It describes the efficacy of local knowledge collected and presented using new GIS technologies in providing alternative management priorities and strategies.

Introduction

This case study tells the story of a maritime people who have depended for many generations on the sea and its fishery resources for their livelihood. In recent years, these people have struggled against government policies which have attempted, and have been largely successful in selling off these resources to the highest bidder. As a result, centuries of fishing tradition and community access to the sea and the survival of the communities themselves are gravely threatened. The aboriginal people of the Bay of Fundy region of Canada have, since time immemorial, lived off the rich marine resources there. In the Seventeenth Century, these First Nations of Canada, including the Mi'kmaq, the Maliseet and the Passamaquoddy, shared their fishing technologies with the new European settlers from France known as the Acadians. After the Acadian settlers were expelled from the fertile farming areas by the British in the Eighteenth Century, the Acadians turned more than ever to a maritime existence. Soon joined by the Loyalists fleeing the American Revolution, these French and English settlers along with the First Nations formed the traditional fishing communities of the Bay of Fundy. In the Nineteenth Century, the Bay of Fundy region boomed with shipbuilding, cod fishing, and trade with Britain, New England, and the Caribbean. Hundreds of tall ships filled the harbours of the Bay of Fundy. With the decline of the offshore salt cod fishery and the "Golden Age of Sail" at the end of the Nineteenth Century, the small fishing communities scattered around the Bay of Fundy consolidated their survival strategies around earning a livelihood from fishing lobster, groundfish, herring and scallops in small boats relatively close to shore. These communities, despite advances in technology and the modernization that followed World War II, persisted as multispecies, inshore fishing communities with a high degree of close social relations and an egalitarian ethic almost to the end of the Twentieth Century. However it was the descendants of the European settlers who were now known as the traditional fishing community in the Bay of Fundy. The First Nation fishing communities after decades of racism and systematic exclusion had become poor, invisible, marginal communities living on reserves with virtually no access to either subsistence or commercial fishing.

Thus in the last decade of the Twentieth Century, the Bay of Fundy fisheries, like so many other fisheries in Atlantic Canada at that time, came to a cross roads. The fisheries of Atlantic Canada were now worth over one billion dollars in landed value and nearly two billion dollars in total value and supported 1,300 small fishing communities representing about one-quarter of the total population of Atlantic Canada. Since 1982, the federal government has instituted a market-based approach to resource allocation in the Canadian fishery through individual transferable quotas (ITQs). The government started with the large corporately owned offshore fleets and by the beginning of the 1990s had finally turned their attention to the vast majority of fishers that operated the small boats of the inshore fleet. With ITQs, fish harvesters had to buy access into what had always been a public resource. And as with any valuable real estate, the high cost of access into a privatized fishery was beyond the financial means of many family fishing enterprises. The inshore fishers of the Bay of Fundy knew that ITQs would spell the end of their traditional way of life and they decided to challenge government policy by offering an alternative: community-based management. At the same time, the people of the First Nations communities who were by no means strangers to the notion of community-based management began to demand social justice. Thus two battles emerged within a relatively short time frame: the struggle of the descendants of the European settlers to hold on to their traditional communities and access to the sea and the struggle of the First Nations to regain the traditional way of life that had been taken from them some time ago. Would these two groups become allies or opponents? Would they combine their efforts towards community-based management or engage in a competitive battle for a dwindling fishery resource?

The Initial Crisis: The Non-Native Fixed Gear Commercial Groundfishery

The starting point of this case study is 1994. At that time, the fixed gear groundfishery was the single largest management unit in the Scotia-Fundy region which stretches along the entire Atlantic Coast of Nova Scotia and around the Bay of Fundy to southwest New Brunswick. There were over 3,000 fishers licensed to use longlines, handlines, and gillnets, primarily in small inshore vessels less than 45 feet in length to catch cod, haddock, pollock, cusk, halibut, flounders, and hake (see photo of fishing boat). In this year, this fishery reached a crisis situation that had two components. The first was a resource crisis. Cod stocks in the Atlantic had either collapsed or were in severe decline. Quotas were cut by as much as 60% and this had followed a cut of 42% in the previous year. In addition, quotas for haddock, pollock, and halibut were half or less of what they had been during the previous decade.

The second was a management crisis. The federal Department of Fisheries and Oceans (DFO) announced that it would restrict its management activities to a core mandate of conservation and would thus no longer manage the fishery to achieve "social" objectives. In practical terms, this meant that the DFO would no longer enforce trip limits; a limit imposed on the weekly catch of each vessel to ensure that catches were equitably distributed among active participants on a seasonal and regional basis. Without this management measure, the meagre quotas could be quickly gobbled up by the largest vessels at the beginning of the fishing year. As a result of the resource and management crises, many fishers and their communities faced a total loss of income from the groundfish component of their fishing activities.

The Establishment of Community-Based Management in the Fixed Gear Groundfishery

In response to this situation, the leaders of the fish harvester organizations, together with the DFO, held a series of workshops and meetings. The result was the development of 18 different management plans reflecting the various socioeconomic situations of different gear types and communities. These plans were approved by the DFO, but now it was up to the fishers to implement and enforce the plan during the 1995 fishing season. Since there was no legal basis for fisher management, the fishing organizations had to get each licence holder to sign a legally-binding contract to obey the management plan developed, implemented and enforced by the fishing organizations.

Despite chaos and confusion, the 1995 fishing season provided fishers with the conceptual and practical experience of developing the notion of community-based management. Thus for the 1996 fishing season, the fisher organizations developed a system of community quotas. The total allowable catch allocated to the fixed gear sector would be divided up into seven portions, based on historical catch rates, to be managed by the licence holders residing in a particular cluster of communities. Each of these community clusters formed a management board that, through a democratic process, devised a management plan for each of the gear types fishing under its jurisdiction.

However, the federal government was committed to ITQs, a market-driven allocation process based on the private sale or lease of quota between licence holders. It appeared that certain DFO officials were undermining the plan for community-based management by negotiating an ITQ system with some of the fishers with the highest catches. If successful, these fishers would carve out a significant portion of the total quota for themselves, leaving the community-based boards with far less quota and dividing the fleet between community-based and market-based allocation regimes. A group of a half a dozen fishers were so outraged by this possibility that they occupied one local office of the DFO, blocking the employees' access to the workplace. Soon occupations spread to other areas until the majority of offices in the Scotia-Fundy region were occupied and non-functioning. At the same time, thousands of fishers, their families, and their neighbours were marching outside the DFO offices, demanding negotiations for the establishment of community-based management and a moratorium on ITQs. The demonstrations and occupations became headline news throughout the country for several weeks. Eventually the federal government conceded and began negotiations for the implementation of community-based management. For the 1996 fishing season, 5 community management boards were established, by 1997 there were 7, and by 1999 there were 10.

In the Bay of Fundy region, there are two management boards, the Fundy Fixed Gear Council and the Southwest New Brunswick Fixed Gear Groundfish Management Board. Both are structured to give the primary institutional role to fish harvester organisations. Each year, quotas for cod, haddock, and pollock are allocated to the boards, based on their collective catch history. In consultation with their members, each organization participates in the development of a management plan that is consistent with local fishing patterns, the traditional distribution of catches among gear-types, local ecological conditions, and community needs. Decision-making attempts to include all licence-holders, and the achievement of consensus is mandatory. In both boards, gear-specific fisher committees make in-season adjustments to the management plans, in response to real-time developments in the fishery, and communicate decisions effectively to fishery participants through a mandatory hail-out/hail-in system. Since the boards have

no formal legislative capacity to enforce their management plans, enforcement is achieved through contract law. Each board requires fishers to sign a contract that they will obey the management plan and accept penalties for violating the plan as decided by an "infractions committee." On both Bay of Fundy boards, the infractions committees maintain the anonymity of both the committee members and the violators. Those fishers who refuse to sign contracts or to accept sanctions can fish under a DFO generic management plan. The fact that only about 1% of licence holders do not sign contracts attests to how much fishers would rather manage themselves than be managed by the government.

The Outcome of Community-Based Management in the Fixed Gear Groundfishery

Despite the success of community-based management in the fixed gear groundfishery, the federal government of Canada remains firmly committed to market-based resource allocation through ITQ's. Other than some administrative assistance to the existing groundfish management boards, the government has provided no support for community-based management. In fact, instead of creating the policy, legislative, and financial framework that would strengthen and expand community-based management, government continues to download a host of its own management costs on to the fishers in various forms of licensing and user fees. The result is that community-based management is threatened with crushing costs associated with paying for two systems of management. They pay for their own community-based system and also for the government's system which is predicated on a small number of lucrative enterprises that financially support the current administrative and scientific management structure. Thus the groundfish harvesters still face a dual crisis. Groundfish stocks have failed to recover, despite stringent conservation methods, and government policies still favour a corporate, rather than community oriented form of development. The difference now is that the fixed gear groundfish harvesters have devised an alternative which has allowed them to survive to this point and demonstrates that there is another way of managing fish that can achieve many of the same economic objectives promoted by the government but with more equitable social outcomes. There are still other outcomes of community-based resource management that must be noted.

Among these are an expanded notion of community beyond the "community of interest" represented by fish licence holders, a greater engagement in scientific research and conservation, and the development of an integrated set of principles that respect the social and ecological community. The Bay of Fundy Marine Resource Centre The fishers belonging to the Fundy Fixed Gear Council on the Nova Scotia side of the Bay of Fundy realized from their struggles that community-based management would have to reach deeper into the community in order to ensure its long-term survival. Therefore, they initiated the Bay of Fundy Marine Resource Centre (MRC), which would provide facilitation, research, and support services to all marine industries. Once initiated, the fishers immediately handed over responsibility for the establishment of this centre to a board of directors comprised of community members.

Although still oriented toward marine industries and benefiting from fisher participation, the MRC now plays a leading role in many community-based development projects in the Nova Scotia coastal region of the Bay of Fundy. It has also inspired the establishment of a number of marine resource centres which are being set up in other areas of Nova Scotia and New Brunswick. Local Knowledge and Conservation.

The empowerment experienced by fish harvesters in the establishment of community-based management helped them to appreciate the value of their own knowledge. The management boards, through their fishing associations, the MRC, and the newly established Centre for Community-Based Resource Management (CCBRM) at St. Francis Xavier University, launched a number of research projects. The impetus for these projects and their research questions were driven by the fish harvesters, rather than by professional scientists or government officials. For example, fishers had long argued that cod stocks consisted of numerous local stocks with each one needing special protection. The government scientists held, on the other hand, that the cod in the Bay of Fundy were part of one large "cosmopolitan" stock that extended out to offshore areas far from the Bay.

Therefore, one global quota was sufficient to protect this stock. Led by the CCBRM and the fishing associations, groundfish harvesters were interviewed and their knowledge plotted through GIS mapping. The study not only provided strong evidence for the existence of local stocks but also graphically demonstrated how much groundfish have been overfished. For example, Maps 2-5 show how the spawning areas of haddock have shrunk during the past fifty years. As a result of this research, the fish harvesters can begin to consider ways to restore lost spawning areas, both through fishing effort controls and more general ecosystem approaches.

Writing the Rules

Shortly after the establishment of the groundfish management boards, the fisher associations and unions that developed and operated the boards in the Bay of Fundy area, decided to "scale-up" their efforts to a regional level by forming the Bay of Fundy Fisheries Council. The Council's mandate was to consider issues of common concern to fishers and it soon developed a strong ecosystem approach to fisheries management issues. One of the Council's first efforts was to join with whale conservationists and environmentalists in successfully opposing a proposal by a private company to harvest zooplankton (krill) for the production of feed for the salmon aquaculture industry.

The Council also launched a "Writing the Rules" project in which many "kitchen" meetings were held with fishers in order to develop their own principles for managing the fishery. The principles agreed to by all the fishers are:

Fundamental Principles

I Fishermen Must Hold Authority in Management

II Management Decisions Must be Made at the Most Local Level

Operative Principles

1. Protect Reproduction
2. Protect Juveniles
3. Maintain Food Chain Relationships
4. Maintain Critical Habitat
5. Protect Local Stocks
6. Establish and Recognize Effort Controls
7. Diversify Fishing Livelihood
8. Promote Owner-Operated Fisheries
9. Retain Fish as Public Resource

These principles have since been widely quoted by other supporters of community-based management and incorporated into the operating guidelines in other community-based management experiments.

THE SECOND CRISIS: THE MARSHALL DECISION

The last few pages of this story focused on the commercial groundfish fishers who are largely the descendants of the European settlers arriving in the

Sixteenth Century and onwards. However as that story was proceeding, another one was unfolding.

During the 1990's, Donald Marshall, a Mi'kmaq fisher was arrested by the DFO for commercially fishing eels without a licence. Donald Marshall was already well known for having spent eleven years in jail for a murder he didn't commit. The subsequent Marshall Inquiry had led to significant changes in the Canadian criminal justice system. Marshall pleaded not guilty to the

charges of illegal fishing on the basis that the treaties signed by his people with the British Crown centuries earlier had enshrined his right to fish commercially. Marshall was convicted by the lower courts and took his case to the Supreme Court of Canada where, in 1999, in a landmark decision, Marshall was found not guilty. The Supreme Court decision went on to explain how every Mi'kmaq,

Maliseet, and Passamaquoddy person had an inherent right to fish for a moderate livelihood. This right was to be managed by the First Nation communities. While the First Nations celebrated the fulfillment of social justice, the non-aboriginal commercial fishers began to panic. They feared their own livelihoods, the value of their licences, and the fishery management system itself would be destroyed by a flood of new entrants from the First Nations communities. In the Bay of Fundy region, the situation reached an extremely dangerous level when scores of non-aboriginal commercial fishers went out in their boats with guns to confront any First Nation person who dared to fish.

In an historic meeting between the Chief of the Bear River Band and the commercial non-native fishers, an understanding was reached which avoided the confrontation. Subsequent discussions with the participants in this meeting and a documentary produced by the Canadian Broadcasting Corporation reveals how much of this understanding was achieved through a shared appreciation of the principles supporting community-based management as revealed in the "writing the rules" project.

Dialogue and Hope

By 2001, the DFO had signed agreements with the majority of First Nations communities. The agreements provided the First Nations with funds to build their capacity to participate in the commercial fishery in return for their agreeing to follow a fisheries management plan that was acceptable to the DFO. With a couple of tragic exceptions, the entry of First Nations participants into the commercial fishery followed an orderly process whereby the licence of a non-native fisher was bought by the government and then transferred to a First Nations community. However manyquestions remain about the future of First Nations fisheries. One of the major questions is: Will First Nations develop a truly community-based fishery that supports the long-term sustainability of their community or will they aaopt the government's model which provides significant benefits to just a few individuals and companies?

In a similar vein, through dual challenges of quota allocation and overfishing, inshore fishers have lost access to much of the groundfish, herring, and scallop fisheries. Thus, the inshore fishery and the small communities it supports are surviving on the basis of high catches in lobster fishery. In the Bay of Fundy area, lobster fishing accounts for somewhere between 80 and 100 per cent of the fishing revenues of individual inshore fishing enterprises. If lobster catches should return to their average historic level, the inshore fishery is in danger of disappearing completely. Out of this bleak situation, community-based management remains a beacon of hope for inshore fishing communities. Since the end of their confrontation, the aboriginal and non-aboriginal fishers have commenced a dialogue about their beliefs, principles, fishing methods, and aspirations. The establishment of community-based management in the groundfishery put in place the social institutions necessary to foster such a dialogue. Led by the MRC, with the support of such institutions as St. Mary's University and the Centre for Community-Based Resource Management, the aboriginal and non-aboriginal fishers have embarked on a number of joint projects and have been allies in confronting environmental issues.

In the Spring of 2003, non-native fishers from the Lobster Advisory Committee approached the DFO advocating the initiation of an experimental lobster fishery by the Bear River Band, one of the few bands that has so far refused to sign a comprehensive agreement with the DFO. In the late Summer of 2003, fishers, native and non-native, from the Bay of Fundy region, climbed aboard a bus and for 10 days visited coastal communities in Maine and Massachusetts to conduct a series of workshops. At the workshops, press conferences and media interviews during this tour, representatives of the two groups stood together and spoke of a future fishery based on community-based management. At the last public meeting of the tour held at the New England Aquarium on the Boston waterfront, one of the organizers of the tour, Arthur Bull, summed up the experience in these words:

We've learned that ways of doing community-based management are very diverse. Each community has its own unique situation, and practitioners of community-based management have to adapt their strategies to each of those unique situations. But, despite this diversity, there are common themes and principles we all share. One is that we must all take responsibility for our future, and steward our resources so they are there for our grandchildren. Also, for community-based management to be successful, we have to develop the capacity to do our science locally, and to change public policy in the public interest.

Lessons Learned

In expanding upon the above citation, there are two lessons that stand out in the experience of the Bay of Fundy fishers. The first is the importance of inclusivity. In 1994, the fishers came to the realization that they needed to stop competing with each other and start cooperating if they were going to survive the dual crisis in the groundfishery. Rather quickly, they learned that this cooperation must not only be with the fishers in their own village community but with the community residents in general and also with the community of fishers throughout the Bay of Fundy. Through articulating a set of principles, they also came to the conclusion that one group among them, the First Nations people, who had been systematically excluded for centuries from their historic livelihoods, had to be part of this ever expanding notion of community. In the end, we saw that this notion of communityexpanded beyond the borders of their own nation and reached out to the fishing communities of New England.

The second lesson learned is the integral connection between fishers, the human community, and the wider ecological community in which they live. The mutual dependence of the groundfish harvesters on one another as fellow human beings was connected with their mutual dependence on all beings in the ecosystem. This human-ecological relationship is strongly reflected in the principles they developed. They discovered that these principles were not only shared by their First Nation neighbours but also that these neighbours had much to teach them about this approach. And despite the dire prospects of the groundfish and First Nations fishers in terms of their current participation in the fishery, they stand out as the best prospects for building a future fishery based on the sustainability of both the human and ecological communities.

INSTITUTIONS, ECONOMICS AND CONFLICTS: FISHERIES MANAGEMENT UNDER PRESSURE

Introduction

Poorly defined property rights, increasing effort and decreasing catches are resulting in conflict in fisheries throughout the world. While these effects are common and well documented in highly regulated fisheries in the European Union (conflicts over allocation of quota for example) and North America (for example conflicts over access to salmon stocks between the USA and Canada); the effects upon fisheries in developing countries, while under-researched are invariably more dramatic and can have long-term implications for the promotion of sustainable livelihoods. While there is as yet little evidence of a sharp increase in natural resource conflicts (Hussein, 1999) it is apparent, however, that the consequences of conflict in the management of natural resources are becoming more detrimental to long-term sustainable exploitation and are effecting an increasingly large number of people (Olomola, 1998; Homer-Dixon, 1994; Streiffeler, nd; Myers, 1987). Although there is an extensive literature on natural resource conflicts, little work has been done on a) analysing the causes of conflict beyond the case-study arena or b) applying economic tools to the study of conflict. In an attempt to redress the balance, this paper sets out some initial findings from an on-going study into the management of conflict in tropical fisheries. Drawing on New Institutional Economics and common property resource management theory, it analyses how and why fisheries institutions adapt to changing circumstances and the role of conflict in the process. Using evidence from Ghana it examines the emergence of fisheries management institutions under differing access regimes and analyses the factors which appear to have influenced institutional change.

The paper argues that conflict can be the result of rising transaction costs and the inability of natural resource institutions to manage these changes. It also argues that conflict can be both a positive and negative force and should not necessarily be eliminated altogether. The paper is divided into four sections. First, it locates fishing as an economic activity within developing countries and looks at the interactions between local, national and international policy objectives and how these can impact upon the development and exploitation of the resource. In the second section, conflicts are described and why they emerge in natural resources management is discussed. Focusing on the role of institutions it looks at how changes in institutions might affect the management of common property resources. Section three draws on initial findings from field research conducted in Ghana between March and May 2000. In

the final section, the paper presents a number of conclusions as to the process of conflict formation and the possibilities for managing it. The role of fisheries in developing economies Fisheries play an important role in the local and national economies of many developing countries where they provide food, income and employment for large sections of the population. However, the framework within which they operate is often complex and vulnerable to failure. The absence of clearly defined property rights, the trade-offs between national and local needs and the impact of rising pressures on the economy can, however, all contribute to the failure of fisheries to positively contribute to the development process.

The national perspective: Following the second world war the rapid expansion of fishing capacity in many developing countries was promoted through the so-called Blue Revolution.

New fishing technologies and fishing methods coupled with a desire to improve nutrition and generate foreign exchange led many countries to build on fishing activity already present. Some of this development was as part of national development programmes, in other cases, international loans and funding programmes helped promote the development of national fishing industries. As a direct result of this drive to build up national fishing capacity, the participation of developing countries in global fishing activity rose dramatically between 1955 and the 1980s. The rewards to be gained from participating in the global fishing market were potentially large, and many benefited, albeit in the short-term, from their forays into fishing.

Increased export-revenues and the multiplier effects associated with expanding industries all helped boost the development of many developing coastal states.

The long-term consequences of such developments, have, however, had a dramatic effect. Rising incomes and increased borrowing of cheap petro-dollars contributed to chronic debt problems as the global economy staggered to a halt in the early 1980s; an attitude of positive speculation on fish-stocks, lack of proper assessment and regulation led to the over-exploitation of a number of important stocks (Schurman, 1996; Thorpe et al 2000).

As a consequence of the international debt burden and the strictures placed on the economy by the Structural Adjustment Programmes, trade-offs between local and national needs put further pressure on resources. The downsizing of the state and the increased role of the market as the resource allocation mechanism saw many developing countries expand their export fisheries to generate revenue whilst at the same time the power to regulate such activity was hampered bythe limited capacity of Fisheries Departments to undertake interventions. The push for export revenue and the renting of grounds to Distant Water Fleets has resulted in spatial conflicts between industrial and artisanal fleets (Johnson, 1996), state support for local fisheries has been constricted and external economic pressures now drive national policy formation (see for example the work by Stonich, Bort and Ovares, (1997) and Neiland, Soley and Baron (1997) on the expansion of shrimp farming)

From a local perspective: In agricultural-based economies, natural resources form the social safety-net that supports large parts of the population. Access to common property natural resources such as timber, fuelwood, grazing land, fisheries, forest products and irrigation water are fundamental to the livelihoods of many of the world's poor. These natural resources are often held as common property and the survival of rural populations has traditionally relied upon a complex set of institutional arrangements that govern use of and access to these resources. The formation of institutional access and use arrangements is well-documented (eg.Ostrom, 1994; Alegret, 1996; Feeny, 1988). Whilst many communities have succeeded in maintaining their common property resources through collective action, in some cases pressures internal and external to the community have caused arrangements to collapse (Berkes,

1986; Mensah-Abrampa, 1988; Adam, 1994). The outcome of a resource becoming open-access is likely to be over-exploitation and the dissipation of resource rents (see Cunningham et al, 1995 for example). In terms of development, then, as the ability of a community of users to exercise rights over their resource is eroded so the role of the resource as a safety-net is undermined.

The pressures that lead to the erosion and collapse of common property resources are diverse. They can result from a rapid rise in population (either through natural growth or an influx of migrants); changes to the ecological status quo (through droughts or flooding, for example); national policy changes that affect rural employment rates; technological change that results in higher harvesting potential and social change that can alter income differentials, systems of property rights and market systems. An important outcome of the breakdown of common property arrangements is the emergence of conflict between stakeholders. However the precise mechanism is not fully understood. This will be explored in the following section. An anatomy of conflict

A review of the conflict literature reveals numerous theoretical approaches to describing and explaining conflict. Conflict is a function of social structure (sociology), of power or class relations (political science) and of individual utility maximisation (economics). It is both positive and negative, constructive and destructive (Powelson, 1972), violent, coercive and non-violent (Wallace, 1993). Thisdiversity of approaches reflects the wide range of disciplines that have addressed the subject. There are many possible definitions of conflict, the following synthesis of definitions is used in this paper:

Conflict is a situation of non-cooperation that involves groups of people with differing goals and objectives and yet Conflict is dynamic and can be a positive catalyst for change Although the term 'conflict' often has negative values attached to it, this is not necessarily the case (see for example Warner and Jones, 1998). Conflict can be described as negative when the outcomes are a zero-sum game (no-one benefits) rather than a positive sum-game (everyone benefits) (Powelson, 1972) or where there is a deadweight loss of social resources as a result of the 'guns vs butter' argument (Neary, 1997:493).

However, it is also important to bear in mind that conflict can be positive and attempts should never be made to eradicate it completely or to prevent it emerging. Conflict encourages goods to be produced more cheaply, government to become more efficient, flaws in the set-up of institutions to be ironed out and allows society to function efficiently by resolving small conflicts often (Powelson, 1972). At this point, the question of the difference between conflict and competition arises. Competition implies the existence of rules, conflict arises when those rules are, for whatever reason, disbanded or ignored. In order to help clarify whether conflict is positive or negative it is often helpful to look at what the conflict is over and how fundamental that disagreement is to the social status quo.

Aubert (1963), Boulding (1966) and Powelson (1972) all differentiate between conflicts that are 'within consensus' and those that are 'over consensus'. In the former case the parties to the conflict agree about the value of what they seek but not the means of achieving it. In the latter case the parties to the conflict are unable to agree on the value of what they seek nor on how to achieve it. The potential impact of conflict is thus dictated by the degree of consensual framework within which they are contested and the degree of conflict over basic consensus (Coser, 1972:73).

An economic explanation for conflict

Economic theory provides two reasons why conflicts emerge over natural resources:

a) the allocation of scarce resources requires trade-offs which become increasingly difficult as the demand for and supply of resources changes and

b) short-term personal gain wins out over long-term social needs or benefits.

The resource allocation issue: Natural resource conflicts occur when the resource in question has become so scarce or degraded as to raise issues of allocation amongst the community of users. Under perfect environmental conditions, as the ratio of users to resource grows so expansion takes place: extra land is brought into cultivation, new areas of forest are exploited, different species are fished or fishery activity moves along the coast or further out to sea. Powelson (1972:33) argues that so long as the answer to 'who gets how much' is resolved by producing more, conflicts over allocation are positive because they encourage growth.

When the ecological boundaries of the resource have been reached, further expansion is no longer possible (without adverse consequences) and the division of the metaphorical cake has to change because the option of increasing the size of the cake is no longer possible. At this point, resource allocation mechanisms come under increased pressure to satisfy all users. The equity of resource allocation tends to diminish as the ability to expand production decreases. As government departments attempt to placate

all the stakeholders, so resource allocation decisions have to change and conflict ensues as certain stakeholders gain precedence over others. This is of course nothing new. The trade-offs between equity and efficiency occur in all markets and are faced by all governments. In the developing world however, where democracy may be weak and where powerful interest groups are able to co-opt and unduly influence the decision making process, managing the trade-offs is difficult.

Short-term gain vs social needs: There is a large body of literature that uses game theory and behaviour modelling to explain why common property management institutions (in particular) can fail (see for example: Ostrom, 1994; Ostrom et al 1994; or Walton, 1998). The Nash Equilibrium and the Prisoner's Dilemma illustrate this point neatly: in the classic decision making model-the Prisoner's Dilemma-the best possible move for player A given the best possible move for player B is to not co-operate.

Thus the socially optimal outcome of the 'game' is not a Nash Equilibrium (Gibbons, 1992; (Jennings, 1999 pers. comm; Parker and King, 1995; Varian, 1990). McKean (1992:248) notes, in a similar vein that the short term benefits of cooperating are often outweighed by the costs.

Consequently a community opts for the least-cost option which is to withhold their contribution to the collective goal. A third explanation for the emergence of conflict has also arisen from the more recent New Institutional Economics school, this sees institutions and transaction costs as a key element.

From a new institutional economics perspective, institutions exist to minimise transaction costs and are a means of transcending the social welfare dilemmas that arise out of individual action and help maximise collective welfare (Bates, 1998:35). They are also the mechanism for dealing with 'gaps' left by market failure in insurance and risk assessment (Bates, 1995). Transaction costs are literally the costs involved in negotiating a transaction and have been described as the economic equivalent of friction in the world of physics (Hubbard, 1997). Transaction costs are the costs associated with gaining information, making decisions and carrying out decisions (Abdullah et al, 1998).

In terms of fisheries, they can be divided into the ex-ante costs of collecting information and making collective decisions in the fishery and the ex-post costs associated with implementing collective decisions. Kuperan et al (1999) argue that transaction costs in fisheries arise, mainly, from the fact that the fishery involve multiple stakeholders and differing objectives and long-term goals. Institutions are important as resource allocation mechanisms-markets and fisheries departments, for example, both have a role to play in allocating scarce resources. What is more, it is now recognised that institutions matter for economic development and recent development research has recognised the need to 'get the institutions right' if effective economic development is to be achieved.

Katterman (1998:294) points out that the OECD has finally recognised that technical co-operation, the long-time philosophical foundation of development practice, can amount to nothing if the institutions that it relies upon for success are weak. Thus for both successful natural resource management and economic development institutions able to adapt to changing circumstances are needed. There are two explanations about why institutions might fail in the face of change.

The first relates to the supply and demand for institutional change, the second to transaction costs. Thomson, Feeny and Oakerson (1992: 132) and Feeny (1988) discuss the supply and demand model for institutional change: when the demand for institutional change (to capture gains not possible under existing arrangements) outstrips the ability to supply change, failure emerges. They list relative factor or product prices, the size of the market, technological change and fundamental decisions of government as causes of change that lead to what they term 'institutional disequilibrium'. However, they also recognise that change depends on the State's willingness and ability to help new institutions emerge. This is a particularly pertinent issue in the developing world where, as discussed before, state capacity may be underdeveloped. Because in the NIE perspective institutions evolve to minimise transaction costs (through minimising uncertainty)-when it is no longer able to do this effectively its position is weakened and it is increasingly unable to deliver services effectively or be allocatively efficient (Klitgaard, 1998:337). Using the NIE paradigm and the supply and demand argument above it could, therefore, be argued that a rise in transaction costs, over and above the institution's ability to accommodate this change leads to inefficient operation. This rise in transaction costs could be due to development pressures (political and economic), environmental scarcity (perceived or otherwise) and structural problems (political and economic) which increase the costs mentioned earlier: the costs of collective decision making, information gathering and collective operation.

In the case of fisheries management, strong and flexible institutions are required at both the national and local level. Arguably, the more nested the structure, the lower the transaction costs between the top and the bottom of the system and the more efficient the institution in its allocation role.

In many developing countries this may not be the case. Communication between the different layers of fisheries management are frustrated, the legitimacy of the ruling body to assign resources is often missing or contested and political factions and the action of rent-seeking elites influence the vertical relationship. Institutions, conflicts and fishing in Ghana The political and economic context

Ghana, the former British colony of the Gold Coast, became the first African colonial state to gained independence in 1957. It has had a number of military coups, but is set, in December 2000 to become the first African state in which a military ruler has handed power to a civilian government through the ballot box. Unlike many other countries in the region, Ghana has experienced comparatively little unrest with its neighbours or amongst its diverse tribal and linguistic groups, and it is perhaps this aura of stability and peace at a national level which influences conditions at a local level. In the 43 years since independence Ghana has made some remarkable advances in development and has also suffered some serious set-backs.

Despite significant advances made in improving some of the standard development indicators (life expectancy has risen by 4 years since 1995; adult literacy has risen by 44% since 1970), Ghana is currently struggling with the dual effects of economic adjustment measures and external shocks. Economic reforms have seen public sector budgets slashed and the impact in the natural resources sector, to name but one, has been devastating. All departments are under funded, and in the fisheries department, this has lead, in particular to a reduced ability to monitor fisheries effectively and enforce regulations. The need to raise revenue has seen the introduction of VAT at 10% which is set to rise to 12.5% later this year. Despite well intentioned publicity campaigns to educate the public as to the need for VAT, this is unlikely to ease the blow of a significant price rise on all VATable goods-particularly on basic grains that have shown sharp price rises since last year. Coupled to this is the rapidly devaluing national currency which, while making Ghanaian exports cheaper, makes servicing the national debt (currently $580 million) more expensive daily and dramatically increases the costs of vital industrial inputs. Ghana is an oil importer, and recent OPEC price rises have also hit the economy hard. While exports have grown aided by improving terms of trade and export initiatives put in place as a result of economic reforms, these have tended to concentrate on cash crops such as cocoa.

In the context of fisheries, the prevailing economic conditions are having a number of effects. Firstly the day to day costs of fishing are rising. Fuel prices not only impinge on the length of fishing trips, but exchange-rate depreciation has increased the costs of imported inputs such as nets, ropes and outboard motors. The knock-on effects of such price rises are widespread. Education and health at the local level both suffer as more and more income is diverted away from the 'add-on luxuries' into the basic needs for survival. Although things have not reached crisis level as yet, there is evidence of malnutrition in fishing villages and primary schools lack basic teaching materials (supplied by the community). Despite this catalogue of economic problems, some very interesting results emerged from a recent study of the causes and management of conflict in coastal Ghanaian fishing villages.

THE SURVEY

Between March and May 2000 the Centre for the Economics and Management of Aquatic Resources in collaboration with the Ghanaian Department of Fisheries conducted a survey of 64 fishing villages along the

Ghanaian coast (see table 1). This sample represents 33% of the villages listed in the latest Canoe Frame Survey (1995) conducted by the Department of Fisheries. These villages are spread amongst 4 regions. With no information on the variability of the population it was decided to select one third of all the villages in each region on a random basis. In order to gain a complete picture of conflicts within the coastal zone, interviews were also conducted with inshore-vessel owners and a number of industrial fleet operatives.

Ghanaian artisanal canoe communities are organised around the Chief Fisherman. A complex set of procedures has to be undertaken before any interviewing can take place and no interviewing can be conducted with permission from the Chief Fisherman. Consequently we targeted the Chief Fishermen who on most occasions was accompanied by a number of other village elders, usually comprising his right-hand man and the most skilful or successful fisherman in the village. On a number of occasions large groups of women traders and other fishermen were present at the meeting. For reasons of simplicity and speed a number of tools were adapted from the extensive PRA 'toolbox' to help build PISCES (Participatory Institutional Survey and Conflict Evaluation Study) developed specifically for use with artisanal fishing communities to gather information on institutional arrangements.

An Oasis of Calm in a Sea of Troubles?

Theory would suggest that given the economic context within which Ghanaian fisheries are operating, the open-access nature of the resource and the inherent instability of pelagic stocks, conflict should be widespread and the fisheries management institutions under considerable pressure. Initial analysis, however, suggests that this is not the case. At the time of writing data collection was still continuing in Central and Western Regions. The summary results below are therefore preliminary results on regions already completed.

Villages as institutions. From the results so far, Ghanaian coastal villages appear to be stable and differ little in their organisational structure along the entire length of the coast. The village chief is at the top of the hierarchy, the Chief Fishermen, first point of call for all fisheries matters sits beneath the Chief. Although by law the Ghanaian coastal line is open access (to Ghanaian citizens), in practice, a complex set of local laws and a system of reciprocity and responsibility govern inshore waters. 91.7% of villages questioned confirmed that although anyone could fish off the beach, in practice, they must seek permission or announce their intentions to the Chief Fisherman, this includes residents and migrants. The remaining 8.3% stated that anyone could fish off the beach, and did not qualify their answer. No one reported ever having a request to fish denied.

Once permission has been granted by the Chief Fisherman he then becomes responsible for the welfare of the fisherman and his kin. This system of 'responsibility' is not unique to Ghana and acts as a kind of insurance scheme. There are many internal migrant fishermen in Ghana-some are migrants that seasonally move from one area to another, others are migrants that have been settled in a village for generations but, because they are from a different language group, are still classed as migrants. The distinction between natives and migrants to so powerful than in one village in Western Region, there are two separate landing sites: the Fante-line and the Ewe-line highlighting the two distinct groups that have fished from the village for many years.

The institutionalisation of fisheries management. The local institutional structure of Ghanaian fisheries has, on the face of it, changed very little over the past centuries, and it is perhaps this stability and consistency that helps it remain comparatively peaceful. However, economic reforms, as described above have caused many significant negative changes to the national organisation of fishing in the past five years but in some areas have also ironically tended to bolster the local structures rather than threaten them. The most significant positive change has been the introduction of the Community Based Fisheries Management Programme (CBFM) under the auspices of the Fisheries Sub-sector Capacity Building Project. The success of the CBFM would appear to be largely due to the fact that rather than radically changing the status quo, it has strengthened the existing structures. The Fisheries Sub-sector Capacity Building Project is a joint venture between the Government of Ghana and IDA/World Bank. Started in October 1995 its main objectives were to improve the long-term sustainability of Ghanaian fisheries. Within this main objective was also the aim to improve the capacity of the Department of Fisheries, address the issue of lack of an active management regime, weak institutional and legal frameworks for fisheries and a growing financial and resource crisis in the industry. In order for the sustainability of Ghanaian fisheries to be improved management plans were needed and for these to be successful they would need the full approval of local communities. The Project will run until 2001 when the funding will finish and the Government of Ghana will be responsible for seeing its continuation.

As part of the CBFM, committees have been formed in each community and their first task was to draw up a list of by-laws governing fishing activity there. By drawing up their own by-laws and submitting them to the District Assembly for ratification, local norms would be lent weight and legitimacy. This has a two-way function. Firstly the community feels that its laws are valued because they are recognised by a higher authority and secondly, it engenders a sense of trust and liaison between the communities and the District Assemblies. The make-up of the committees varies but as a rule consists of the chief fisherman and representatives of the various stakeholders. Typically this would include the indigenous and migrant fishermen, the fish processors, fishmongers and fishtraders, vessels, gear and engine owners, and a representative from the Fishermen's Service Centre.

During the course of the field work the villages in a number of districts in Central Region were going through the process of ratifying the by-laws which had recently been gazetted. The drawing up of by-laws has allowed existing norms to be institutionalised and has allowed 'best-practice' in fishing to be argued out amongst the fishermen and agreed up in a formal setting. A good example of this is the by-law that bans children from the beach during school hours (recognising that schooling is a very important part of village development) and the by-law that formally bans fishing with dynamite (an extremely contentious issue up to this point).

Conflicts and conflict management. In the regions so far covered, violent conflict between fishermen was very rarely mentioned. Day-to-day squabbles and difficulties are, however, common. Interestingly, the pricing of fish is regarded as a conflict, although it could be argued that this is an essential and integral part of how markets work and is a sign of competition not conflict. However, because the ethos of the survey was for the communities to identify what they considered to be conflicts, we have let it stand. Two other conflicts often attracted heated debate: the incursion of semi-industrial trawlers into water less than 30 metres deep (reported by 25% of villages) and fights among women over access to catches and credit facilities (reported by 29.2% of villages). Although there was scant reference to a rising number of canoes or fishermen related to the presence of conflict, two examples stand out in particular. One was a fishing community in Accra which suffers from its proximity to the city centre and all the urban poverty problems associated with it. Here the Secretary to the Chief Fisherman mentioned that the landing beach was no longer able to cope with the number of canoes and there were many conflicts over landing canoes and off-loading catches. The other was Mumford, a well documented fishing village also reported problems with lack of landing facilities exacerbated by the increased number of canoes present in the village. Mumford's case is discussed in more detail later on. Based on the answers recorded, conflict in Ghanaian coastal artisanal fisheries can be divided into a number of categories: conflict that results from outside influences, conflict that results from the internal allocation of resources and that which could probably be better described as competition. The typology below attempts to categorise the most frequent problems encountered, it should be noted that at the margins some conflicts could happily sit in a number of boxes. Those conflicts caused by outside influences are the hardest to solve because they involve elements beyond the immediate control of the village.

Because violent conflict is almost unheard of, it is perhaps useful to establish a conflict scale for Ghana that better distinguishes between the various degrees of day-to-day squabbles.

Conflict management at the village level is highly structured and organised and it is perhaps a reflection of this that conflicts are few and far between. All villages in the survey reported that any conflict between fishermen is reported first to the Chief Fisherman who then, along with his panel of elders, comes to a decision on the issue. In the case of damage to nets or boats caused by other canoe owners, the culprit is usually made to pay 2/3 of the damages (recognising thus that fishing is a dangerous activity and incidents are rarely deliberate). Although these cases may take a while to resolve-establishing fault and liability-they are not considered to be onerous. The conflicts that do take longer to resolve are those between different types of vessel-trawlers damaging the nets of canoe fishermen cause a frequent problem.

Even when there are a number of witnesses and broad daylight has enabled the collection of the name and registration details of the offending vessel, the cases can take months to resolve. Such cases are reported to the relevant local fisheries office that then deals with them. Trawler owners often deny all responsibility for the damage and communication and transport difficulties between the administrative centres and the villages further lengthens the process.

During discussions other factors arose that, whilst not identified as conflicts by the villagers, were identified as making life in the villages more difficult. Two examples of these are the incidence of erosion and the rising costs of inputs. Sea erosion was frequently mentioned in villages in Greater Accra and Volta region to the east of Tema harbour. Sea erosion in the Bight of Benin is serious; current large scale project funded by international agencies are working to building sea defences to protect the coastline. Local lore suggests that without these sea defences Keta Lagoon (a significant water body close to the Togo border) will disappear into the sea in five years time. The building of Tema harbour was the most frequently cited reason for erosion in the east of the country. The knock-on effects of the Akosombo Dam, built on the Volta River are also cited for disrupting flooding patterns in the river delta (this has resulted in decreasing water levels in lagoons) and the disappearance of certain fish species at the mouth of the river. The rising price of inputs was universally mentioned as one of the hardest issues to contend with. Given that changing economic circumstances affect how communities deal with the day to day running of their lives, it is hardly surprising that there is a perceived link between increased economic hardship and conflict.

SOME TENTATIVE EXPLANATIONS OF CONFLICTS IN GHANA

Ghanaian artisanal fisheries present some interesting dilemmas in terms of conflict analysis and common property resources. Although Ghanaian waters are legally open access, a powerful sense of culture and tradition, shared amongst all groups has maintained a common-property system in the face of an open access regime. There is a profound sense of history and cohesiveness in fishing communities along the Ghanaian coast and this has undoubtedly allowed all communities to be able to control who does and does not fish off the beach. The overwhelming sense in all villages that they were in control of their own destiny is probably also a factor that has allowed common property access rules to operate in these circumstances. The collective will to protect their right to the resource, and the fact that that collective will is recognised and respected by all the villages is probably the key to the success of the management systems here. Although game theory suggests that short-term gain will often win out over longer terms societal needs, this appears not be the case so far in Ghana. Those in the community willing to work towards social benefits still outnumber those looking for personal gain. Rising transaction costs are a significant issue in Ghanaian fisheries management-both at the local and national level. Increasing economic pressures are making the day-to-day life in the communities harder. There is a boom in building projects in the country-particularly related to the tourist industry in the coastal regions. So far, however, only two villages have reported land-based disputes and interestingly both of these are close to the rapidly developing tourist industry under one hour's drive from the capital.

In both cases a rush to secure prime spots of land has exposed the often fuzzy and complicated land rights in the region. While in one village the land dispute was a major issue but totally unconnected to fishing, in another the dispute between two rival villages (both claiming rights over a prime piece of real estate) has found its way out to sea. It was reported that whereas neighbouring villages would previously help each other at sea in the case of accidents, that is no longer the case and the incidence of damage to nets and canoes between the two communities is on the rise. Further research will identify how common a problem this is along the coast, but in the Greater Accra region it is would appear that externally generated conflicts are the ones that put the most pressure on the ability of the community to solve its problems.

Economic policies are also caused the price of inputs to rise. This has a number of knock-on effects in the community. The rising costs of inputs impacts upon conflicts over the price of fish as both buyers and sellers aim to maximise their profits in increasingly difficult circumstances. The cost of fishing rises which, whilst not immediately causing more conflicts, puts added pressure on the day to day existence of fishermen and brings economic pressure to bear on the rest of the community.

Despite the increasing costs of living around them the fisheries management institutions within the communities would appear strong enough to be able to rebuff external shocks.

The rise in transaction costs is also evident in other sectors. An interesting problem that was raised by a number of villages was the lack of landing facilities. One community in Accra reported that there were too many canoes attempting to land on a small beach, and the village of Mumford in Central region reported a similar problem. Here the fishermen are not able to beach their canoes and have two options: leave them out at sea or take them to nearby Takoradi. Given the nature of the sea the latter option is preferred which has a number of impacts on the village. Firstly the transaction costs of the fishermen rise because they are travelling further to land their catch and have to negotiate access with the relevant Chief at the other landing place. Secondly, less fish is consequently landed at Mumford which impacts upon the livelihoods of the women fishmongers there. Travelling to meet the canoes involves transaction costs that, with the current economic climate are rising daily. It also involves the less obvious transaction costs of negotiating with fishmongers at the new landing site who now have increased competition in the purchase of fish. Once again, it would appear that the strength of the village communities and the mutual respect amongst the neighbouring villages has led such incidences to be considered as minor disruptions to daily and, so far, are not the source of any destructive conflicts. The intrusion of trawlers into coastal waters is another factor that has seen transaction costs rise. Reporting the incident takes time and effort, resolving the issue requires even more. Travel and communication between many of the very isolated villages and the district headquarters is costly and time-consuming. The trawler owners are often based in ports far from the site of the incident and are reluctant to admit liability and often deny all charges. Despite mechanisms to resolve such conflicts, they often remain the most costly conflicts in terms of time and money. As the incidence of trawler incursions rises, so presumably will the transaction costs associated with this.

Only increased funding of a monitoring and enforcement capacity and negotiation with the trawler owners is going to be able to better deal with this problem. This is supposed to be covered under the remit of the FSCBP, but as it liable to be a long-term goal. The root cause of the trawler issue-declining catches and rising costs pushing the trawlers into illegal areas-has a number of long-term solutions.

However, the trawler sector is facing severe economic pressures and any measures are probably too painful to contemplate at this juncture and beyond the capacity of the State (economically and politically) to implement currently. The Community Based Fisheries Management Programme (CBFM) would appear to have had positive benefits to the communities. By using the existing institutional structure, the CBFM has enabled more formalised management to be introduced to communities without upsetting what was clearly a system that worked well to begin with. The only communities that appeared indifferent to the CBFM were those in urban areas (principally Accra). The reason given for this attitude is that fishermen in these communities tend to be better educated and have greater access to media and other information. They are more cynical of government promises of how things will get better in the future and more likely to rebuff any attempts by the government to interfere. Outside urban areas, the CBFM was the reason cited for the decrease in conflict or the satisfaction with the conflict management system in place. The gazetting of local by-laws was looked upon favourably by all those spoken to. So, in terms of transaction cost analysis, it could be argued that providing local institutions with a more 'legitimate' basis both within their own community and in the district as whole has helped maintain transaction costs at a stable level. In those communities where the incidence of conflict has declined, the CBFM may even have helped reduce transaction costs. There is, however, a corollary to this argument. Although the CBFM has had a number of positive benefits, it has also inserted bureaucratic systems into institutions that worked quite happily without them before. Further research needs to be done on this issue, but it would be interesting to see if CBFM has actually increased the transaction costs of some communities in terms of time spent at CBFM meetings, costs incurred in travel to meetings and the added costs of bureaucracy involved.

Conclusions

Conflicts arise in all fisheries regardless of size or structure. However, where the community has a degree of control and immediate interest in the health of the fish resource, there is more likely to be a collective will to manage the resource for the benefit of the community as a whole and to thus mitigate the effects of conflicts. As control of the resource is removed from the community of users, so the incentive to act collectively diminishes and the conflicts erode any sense of 'rules of engagement'. Although much has been written about natural resource conflicts, little is understood about how these conflicts emerge, develop and are managed. In an attempt to improve the understanding of conflicts, this research examines the role of institutions in the process. In particular, it is interested in the contribution that economics can make to the understanding of institutional capacity and conflict formation. Although physical factors often spark off conflicts: a rise in the number of resource users, the decrease in fish stocks, climate and ecological changes, it is the ability of institutions: both formal and informal to manage this change that is critical. In the case of Ghana, it would appear that strong local institutions, supported, albeit at a minimal level by the state, have helped maintain conflict at a low level, despite the rising transaction costs around them. A number of issues remain beyond the realm of influence of local institutions-sea erosion and trawler activity in particular. However, the fisheries management mechanism in Ghanaian coastal communities is strong enough to be able to absorb rising transaction costs and internalise them thus keeping conflicts to a minimum.

Bibliography

Beckham, Stephen Dow: *The Indians of Western Oregon: This Land was Theirs*, Coos Bay, Oregon: Arago Books, 1977.

Bergin, A.: *The Pacific Patrol Boat Project: A Case Study in Australian Defence Cooperation*, Australian Defence Force Academy, Canberra, 1994.

Bigart, Robert and Clarence Woodcock: *In the Name of the Salish and Kootenai Nation: The 1855 Hell Gate Treaty and the Origin of the Flathead Indian Reservation*, Pablo, Montana: Salish Kootenai College Press, 1996.

Boyd, Robert: *People of The Dalles: The Indians of Wascopam Mission*, Lincoln, University of Nebraska Press, 1996.

Bullard, Oral: *Crisis on the Columbia*, Portland, Touchstone Press, 1968.

Cohen, Fay: *Treaties on Trial: The Continuing Controversy over Northwest Indian Fishing Rights*, Seattle. University of Washington Press, 1986.

Cone, Joseph: *A Common Fate: Endangered Salmon and the People of the Pacific Northwest*, Corvallis, Oregon State University Press, 1995.

Crane, Barry D. and Steve Warner, Melissa Hirsch Kuchma: *Fisheries Law Enforcement: Assessment of Deterrence*, Institute for Defense Analyses, Virginia, 1999.

Kelleher, K. and P. Cacaud: *Governance of Yemen's Fisheries: Analysis and Project Financing Proposal*, Prepared for the European Commission, Tebodin, 1999.

Landeen, Dan and Allen Pickham: *Salmon and His People: Fish and Fishing in Nez Perce Culture,* Lewiston, Idaho: Confluence Press, 1999.

Marohn, E.J.: *Cost Effectiveness Relationship: Operational Organization of Monitoring and Optimization of Tactics. International Conference on Fisheries Monitoring, Surveillance and Control*, Brussels, Belgium, 1999.

Peterson, Keith: *River of Life, Channel of Death: Fish and Dams on the Lower Snake*, Lewiston, Idaho: Confluence Press, 1995.

Richards, A.: *Reported Fisheries Violations in the EEZs of FFA Member Countries, 1978 - 2001: An Analysis of the Violations and Prosecutions Database*, Forum Fisheries Agency, 2001.

Taylor, Joseph: *Making Salmon: An Environmental History of the Northwest Fisheries Crisis*, Seattle, University of Washington Press, 1999.

Ulrich, Roberta: *Empty Nets: Indians, Dams, and the Columbia River*, Corvallis, Oregon State University Press, 1999.

Index

A

B

C

□□□